Email and the Everyday

Email and the Everyday

Stories of Disclosure, Trust, and Digital Labor

Esther Milne

The MIT Press
Cambridge, Massachusetts
London, England

This book was set in Stone Serif and Stone Sans by Westchester Publishing Services.

Library of Congress Cataloging-in-Publication Data

Names: Milne, Esther, author.
Title: Email and the everyday : stories of disclosure, trust, and digital labor / Esther Milne.
Description: Cambridge, Massachusetts : The MIT Press, [2021] | Includes bibliographical references and index.
Identifiers: LCCN 2020016674 | ISBN 9780262045636 (hardcover)
ISBN 9780262552660 (paperback)
Subjects: LCSH: Electronic mail messages. | Electronic mail systems. | Language and the Internet.
Classification: LCC HE7551 .M55 2021 | DDC 384.3/4--dc23
LC record available at https://lccn.loc.gov/2020016674

Experiencing the everyday can sometimes be a gift. I realized this fact when my mother died. She had offered a rock-solid foundation of normalcy, comfort, and the deep security of familiarity. Like many wonderful aspects of life, I didn't quite grasp the significance until it was gone. I dedicate this book to my parents, Ruth and Rex Boschen, with love and beautiful memories.

Contents

Acknowledgments

This book would not exist without the contributions of my research participants. Thank you for your funny, sad, worrying, and intimate stories of email. I hope in these pages you find little sparks of recognition and enjoyment. Many thanks to Kaz Horsley for her meticulous editorial work. For his patience and assistance with the survey design and analysis, my thanks to Scott Ewing; you are so very much missed. Mentoring is a word often bandied about, but I need to thank Karen Farquharson, Larissa Hjorth, and Alison Young for just this sort of support over the years. I thank the anonymous reviewers whose advice and astute suggestions helped me navigate the path to publication. And, finally, thanks to the Special Interest Group for Computers, Information and Society (SIGCIS) for their knowledge and generous sharing of vital details. In particular, my warm thanks to Thomas Haigh for his insightful reading of chapter drafts.

This book was in production when the COVID-19 pandemic struck in 2020, and so its relation to email has not been considered.

Introduction

In April 2014 a story began circulating across social networking sites announcing the ban of email. Apparently France had called a halt to the use of employee email communication after 6 o'clock at night.[1] The move was the result of a labor agreement struck between employers' federations and workers' unions in the engineering and IT sectors (including "the French arms" of Google and Facebook and some high-powered consultancy firms). It stipulated that staffers in these companies had the right to disconnect from electronic communication devices to ensure their working time did not exceed the legislatively mandated 35-hour week. Reactions to this news item were instructive. UK-based media chastised the French for their "quaint ideas." As one columnist from *The Independent* elaborated:

> Heaven only knows what the average British working week would be if digital hours were taken into consideration. No matter what time of the day or night, whatever we may be doing in our leisure hours, we are only a ping away from being back at a virtual desk. I rarely have dinner with anyone these days who isn't attached to their smartphone, waiting for a pause in the conversation so they can check their emails.[2]

Similarly, a *Guardian* columnist joked, "While we're staring down the barrel of another late one ... across the Channel they're sipping Sancerre."[3] US news reactions echoed this incredulity with a *New York Magazine* writer sighing, "Well, at least we American workhorses can still end a bad date with the ol 'Oh, sorry, my boss is emailing me like crazy!' exit strategy."[4] Although these articles ostensibly argue for achieving sustainable "work-life balances," they somehow reinforce the imperative to respond always to the ceaseless demands of the market. With no such gesture toward labor reform, *USA Today* called the move "draconian,"[5] and a Fox Business policy analyst found the agreement "absurd."[6]

When the French "right to disconnect" law was enacted in January 2017 it attracted a fresh round of interest about the "always on" digital workplace. Again, while there was widespread approval about new ways to achieve a reasonable balance of work and leisure hours, running beneath some of the reports was a slight distrust for those who would question their workplace labor entitlements. The disadvantage of such legislation, which was part of a suite of new French labor laws, was an increase in the bureaucratization of everyday life. A *Time* reporter who called it the "dark side" of the reforms warned how "reams of laws and regulations" governing termination clauses and annual leave "have proliferated over the years," adding wistfully that the United States "can only dream of" such entitlements.[7]

This news event introduces some central questions I address in *Email and the Everyday*: How is email experienced, understood, and materially structured as a practice that traverses the domestic and institutional spaces of everyday life? What kinds of stories are told—both *about* email and *through* email? How can we understand email itself as a significant industrial sector? What are the consequences for the public release of large-scale email datasets?

Before explaining my approach to these questions, I want to explore briefly how email has been understood in both popular and scholarly contexts. Invented during the 1970s and coming to widespread adoption by the late 1990s, email has been lauded and condemned. In what follows I outline some of the major trends in communication research about email and also point to the gaps: the intriguing omissions that prompted the writing of this book.

The Death of Email

As is often the case with narrative, our story begins with its own death.[8] Underpinning the French news reports was the conviction that it was time for organizations to "kill-off email."[9] One *Wired* reporter made a similar pronouncement—"email's about to die!"—and quoted Facebook cofounder Dustin Moskovitz to back him up.[10] Yet the demise of email has been predicted for at least two decades with various reasons advanced for the threat. In 1989 John McCarthy speculated that the fax machine would bring about the end of email communication, then in its first decade of widespread use.[11] From the mid-1990s, it was spam—the sending of "unwanted commercial email"[12]—that emerged to herald its undoing. Writing in 1999,

Patrick Flanagan warned that the "extraordinarily enhanced connections" and "excitement of interacting with others" through email were jeopardized by the introduction of spam.[13]

From late 1999, an interrelated narrative thread developed around the idea of "information overload," and in these warnings volume was hampering organizational productivity. While in some instances spam was again cited as the reason people couldn't manage their inboxes, and was therefore indicative of the demise of email,[14] research also looked for other reasons why traffic might be on the rise. Online education, for example, increased email to unmanageable levels for educators.[15] Discussing "the last days of email," another commentator linked overload to flattened organizational hierarchy: "anyone can send one to any number of people on any subject at any time of the day," a situation which means we are "at (if not past) the saturation point with e-mail."[16] In 2011, this narrative reached a crescendo when Thierry Breton, the chairman and CEO of the business technology consultancy Atos, announced the abolition of internal email by 2014.[17] As part of an overall "wellness at work" strategy Atos had embarked upon, this new initiative, called Zero email™, would address the "challenges organizations face as a result of the continuing explosion in data." Instead of email, Atos implemented new collaborative tools for internal communication because "email is on the way out as the best way to run a company and do business."[18]

Finally, during the mid-2000s, the death-of-email narrative became one of media obsolescence. A consumer affairs advocacy publication claimed in 2006 that "instant marketing," through "blogs, social networking sites, and mobile text" meant email is "losing its luster."[19] Likewise, a *Slate* article of 2007 argued that "e-mail is looking obsolete" as Facebook, Twitter, and texting become the dominant mode of communication for the teen market.[20] In 2008, the business publication *Wikinomics* announced "more news on the death of email," writing that "over 95 percent of emails received could have been dealt with more effectively by other media including face to face, video or instant messaging."[21] And in a 2010 blog post the social media business platform Socialcast unsurprisingly declared, "Social networks spur the demise of email in the workplace." Even in a field not known for its restraint, the hyperbole of this article is remarkable and worth quoting in some detail:

> Email may be dead, but it's not going to disappear. Email is like a zombie or a vampire—it's going to hover and haunt us when we least expect it.... A communication shift is happening as users look to blow the top off information silos

> and let knowledge flow freely without the constraints, frustrations and loss in productivity email brings.... Now, instead of pockets of knowledge, employees will have one central nervous system that unifies every piece of an organization's information.... This new-found freedom of better information flow will be the nail in the coffin for email.... We'll begin to thrive as we witness and experience the renaissance of social enterprise communication, moving closer each day to email becoming a burden of the past.[22]

This author's compelling images can be placed within the critical theories of residual media[23] or media archaeology.[24] These approaches do not represent a unified field although they do offer productive ways to negotiate or map the uneasy dynamic that exists between the technological past, present, and future. Such perspectives illustrate that the narratives of media redundancy and revolution are never neutral but, instead, are always advancing patterns of economic and institutional relations; what Charles Acland calls a "magically transformative capitalism, lifted from the constraints of the material" whose power depends upon "masking of the hand of labor."[25]

The views of Tim Young, the author of the Socialcast post quoted above about the autonomous, dematerialized force of information, are shared by many other IT business consultants. Ryan Holmes, for example, the CEO of HootSuite, a social media management system says: "Email is the new pony express—and it's time to put it down.... Email is where good ideas go to die. Brilliant messages race across the Internet at light speed only to end up trapped in an inbox."[26] For these commentators email has a sort of regrettable materiality. It "traps," pressing us down, burdening us with a past we long to escape.

I am not interested here in proving or disproving such claims about email by pointing out that the materiality of "social media" may also exert such constraints, but thinking through a residual media of the present allows me to explain what this book is not. It is not an elegiac rescue mission for email. Like many other institutional drones, I struggle with it too. But that is precisely the point and what gives email its piquancy, its urgency as a media form to study. The moment when communication platforms, applications, or cultures give rise to an excess of affect is when research opportunities are incredibly rich. How email can be simultaneously, banal, overlooked, indispensable, and reviled is one of the central questions I seek to explore in this book.

Prominent Trends in Email Research

Despite these dire predictions, as a global mode of business and personal communication email still outstrips newer technologies of online interaction. The Radicati Group, a market research company, reports that in 2020 over half of the world's population uses email and more than 306 billion emails are exchanged every day. During the next four years, email growth is predicted to continue at a rate of 3 percent to reach more than 4.4 billion email users worldwide by the end of 2024. Moreover, email remains the predominant way to sign up for internet-based services including social networking sites, online shopping, government interfaces, telecommunication providers, and news and entertainment portals.[27] Reporting on the most popular online mobile activities worldwide in 2017, Statista finds that "despite the increasing growth of mobile messenger apps, e-mail has remained the leading online communication channel."[28] Email ranks as the no. 1 online activity for mobile phone users ahead of watching video content or accessing social media sites.[29]

My point here is not to argue for the "dominance" of email over other forms of communication or to make predictions for its future. But it is striking, despite figures that clearly indicate the continuing popularity and everyday reliance upon email, how email seems routinely to be ignored across contemporary landscapes of media study. In the fields of communications, media history, cultural studies, material culture, media archaeology, and digital ethnography, the story of email—its role as a form of technological, institutional, and cultural practice embedded in everyday life—remains to be told. Email history as a key point in the invention of the internet has been explored in detail (as I elaborate below) but in general these investigations use email as a stepping stone to the central platform, application, or device with which they are more concerned. Indeed, the term "email" has become something of a shorthand, used merely to stand in for the wider category of "digital communication" or, going back a few years, "new media." In a given media account we might read a statement that says "conventions have developed about the appropriate language to use in email compared to a Facebook status update." Yet turn to the index and "email" is nowhere to be found.

Across the wider research sector many studies of email exist. Spanning three decades, the existing scholarly literature on email communication is, as one would expect, vast. Searches across major academic databases return thousands of hits for email. As a salient reminder of its ubiquity, in many of these citations the only occurrence of the search term—that is, the reason it has returned a result—is the inclusion of an email address of the author or details about how to contact the journal editors by email. To provide a snapshot of current and historical trends in email research, the major critical themes are mapped below. Instructive, too, are the journals in which the highest proportion of email-focused studies appears. Leading the field in this regard is the journal *Communications of the ACM*. Published in 1990, one of its first papers to consider email communication, titled "Networks, Email and Fax," examines the syntax of the email address and the utility of error messages, warning again that email would soon be replaced by the fax machine.[30]

Flaming and Affect

Of the early investigations on email, one of the most widely cited is Lee Sproull and Sara Kiesler's 1986 paper "Reducing Social Context Cues: Electronic Mail in Organizational Communication."[31] Here, the authors examine how email flattens institutional hierarchies because "messages from superiors and managers looked no different from messages from subordinates and non managers." Lacking social context cues of position or status, organizational email produces uninhibited behavior that includes "profanity, negative affect and typographical energy" (called "flaming") and the use of email for personal communication purposes. Since the publication of Sproull and Kiesler's paper, the "cues filtered out" perspective, together with the terminology surrounding the phenomenon of flaming, has been critiqued.[32] A chief objection is the imprecision of the term: hostility, overexuberance, swearing, loss of inhibitions, superlatives, exultation, racism, and misogyny have all been included as defining features of flaming.[33] Although interest in flaming peaked during the 1990s the debates have been reconfigured around contemporary practices of trolling.[34] Early communications research on affect was often investigated through "social presence theory" or "media richness," which examine the degree to which a given application produces the "presence" of one's interlocutor. Like the flaming research, approaches were critiqued for overlooking contextual factors and for presuming certain qualities were intrinsic to the technology.[35] "Medium choice" continues to inform research trajectories about email and affect, with studies developing instruments to assess the motivations or fears governing media selection—often with the aim of making predictions about the psychological contexts and profiles of users.[36]

Such perspectives raise the issue of technological determinism, the specter of which can be glimpsed hovering in the background of the chapters to follow. In order to redress what seems a gap in media and communications research it is necessary to exert a sustained focus on email, its formal properties, the look of its software, and the stories of its use. But email examined in isolation risks abstraction well beyond any recognizable or lived experience of "the everyday." So my approach attempts to move between finely grained analyses of email stories and consideration of wider social or political currents.

Linguistics, Education, and Letter Writing

With conceptual links to flaming, email has been a prominent topic of research within the disciplines of linguistics and education. Early research attempted to determine how the apparent immediacy and colloquial vocabulary of email resembled speech while also capturing some of the qualities of writing, specifically, its asynchronous basis and its geographical and technological separation of users.[37] Of related sustained scholarly concern have been the challenges of email for education in the contexts of English as a Second Language.[38] The field of linguistics and language has produced a number of monograph-length studies including Carmen Frehner, *Email—SMS—MMS: The Linguistic Creativity of Asynchronous Discourse in the New Media Age*; Naomi S. Baron, *Alphabet to Email: How Written English Evolved and Where It's Heading*; and David Crystal *Internet Linguistics: A Student Guide*.[39]

Email is also conceived as a genre of literature through the field of epistolary communication, letter fiction, and the history of the post office. These studies include contrasts between email and letter writing in works such as Brenda Danet, *Cyberpl@y: Communicating Online*; Sunka Simon, *Mail-Orders: The Fiction of Letters in Postmodern Culture*; and Emma Rooksby, *E-Mail and Ethics: Style and Ethical Relations in Computer-Mediated Communication*.[40] Danet's research represents a prolonged scholarly interest in the registers of online discursive modes, together with the traditional media forms of greeting cards, postcards, and letters.

Writers and artists continue to see the narrative potential of email. *Hey Harry Hey Matilda* (2017), by the photographer Rachel Hulin, is a fictive story about twins told entirely in emails. Interestingly, the book has been billed as the "world's first Instagram novel" because early versions of its text and image were posted to Instagram in the run-up to Hulin securing a publisher. For many of these projects, it is email's mundane nature that recommends it for aesthetic purpose. Because email seems almost universal and unremarkable it offers a surprisingly powerful narrative technique. This storytelling quality holds, as I later discuss, whether we are referring to the formal properties of the email novel or the everyday worlds that unfold on email discussion lists and parenting groups.

Organizational Email and Workplace Surveillance

As mentioned, the interrelated disciplines of organizational behavior and management studies were some of the first to consider the function of institutional email, with a high proportion of the early research focusing on digital literacies.[41] Email monitoring and privacy also represents a persistent critical thread running through the literature. A 1992 article published in *Public Relations Quarterly* notes that the increasing prevalence of email into the workplace has been matched by a significant increase in legal action brought against employers for their intrusive monitoring of staff communications.[42] Similarly, a report on human resource planning in 1999 argues there is an urgent need for policy formation about personal email use in business contexts to protect commercial reputation and minimize legal exposure. It reports on a survey conducted by Elron Software, manufacturers of workplace monitoring programs, which found that 85 per cent of adults send or receive personal emails through company systems, 70 percent have sent or received "adult oriented" emails, and 64 percent have sent or received racist or sexist email content.[43] Given the company's interest in producing evidence of product demand, results that demonstrate high percentages of potentially legally compromising email usage are not surprising. Caveats aside, many studies show that in varying degrees, workplace monitoring is a ubiquitous, institutionalized practice.[44] Regular justifications for an employer's need to monitor its email systems are commercial confidentiality, IP protection, minimizing harassment, and maintaining workplace productivity by restricting personal emailing.[45]

Alongside management studies that endorse (or at least do not overtly criticize) workplace monitoring[46] is a body of literature concerned with the protection of employee privacy rights. A 1996 examination about the introduction of the US Electronic Communications Privacy Act of 1986 details its inability to protect adequately the privacy of workplace emails through several unsuccessful court actions brought by employees.[47] Arguing more forcefully, Thomas Hodson and his colleagues state that, although it is legally sanctioned, workplace surveillance is both unethical and economically unviable.[48]

Such studies provide a critical backdrop for thinking through aspects of everyday workplace email. In particular, I discuss how my research shows that disclosure and trust emerge as areas of significant interest. More broadly, and cutting across both the personal and professional contexts for email use, the unexpected publication of or unauthorized access to someone's messages is an enduring cultural trope about email. While we might be familiar with the leaks and hacks of sensational newspaper headlines, in actual fact the micro disclosures of everyday email life need also to be considered.

Email Marketing and Business Etiquette Manuals

Email marketing manuals make up the largest proportion of books published about email.[49] Also popular are email guides and etiquette manuals; two representative titles are David Shipley's *SEND: Why People Email So Badly and How to Do It Better*, and Jim McCullen's *Control Your Day: A New Approach to Email and Time Management Using Microsoft® Outlook and the Concepts of Getting Things Done®*.[50] As well as the "self-help" business literature there exists a wealth of peer-reviewed material studying the commercial possibilities of email from across consumer behavior, direct mail advertising, electronic shopping, and viral campaign tools.[51]

Individual studies like these piqued my interest about the existence of an email industry. It is easy to plot the success of one particular email marketing approach or to drill down on what analytics tell us about the most popular email clients or software. Who leads the market between Gmail and Microsoft, for example? But what's missing is a big-picture sense of the broad email-provider industry: a loosely connected media landscape comprising data-analytics agencies, email software vendors and developers, internet service providers, and email marketing companies. Although important work has provided a systematic account of the commercial reach of the internet—most notably the 2008 edited collection by William Aspray and Paul E. Ceruzzi, *The Internet and American Business*—what's missing is a helicopter view of all the interlocking parts of what we can call the email industry.

The History of the Internet

Email as a key point in the invention of the internet has been thoroughly detailed.[52] The standard history begins in the United States in the mid-1960s with the development of packet-switching technology within a number of industry, university, and government institutions including (in the United States) the Defense Advanced Research Projects Agency (DARPA), RAND Corporation, the research consultancy Bolt Beranek and Newman Inc. (BBN), and (in the United Kingdom) the National Physical Laboratory (NPL). From these various technical, business, and social initiatives the Advanced Research Projects Agency Network—ARPANET, the precursor to the internet—was built in 1969 and began to connect computers at major US and UK university sites.

During the intervening decades, there have been significant attempts to enrich this narrative by looking at lesser known nodes of internet history to include under-researched platforms, applications, and cultural stories across global networks.[53] A special issue of the journal *Media International Australia* includes the use of microcomputers in the 1980s,[54] the rise of political blogs,[55] and the uptake and gendered daily experience of the internet in homes during the 1990s.[56] Finally, *The Routledge Companion to Global Internet Histories* demonstrates the strength of perspectives that investigate internet development outside of the dominant Anglophone approach.[57] Two important monograph works represent productive historical approaches: *The Spam Book: On Viruses, Porn, and Other Anomalies from the Dark Side of Digital Culture*, edited by Jussi Parikka and Tony D. Sampson, and Finn Brunton's, *Spam: A Shadow History of the Internet.*[58] These provide valuable corrective versions of internet history that dismiss spam as an aberration, completely distinct from the authentic story of online culture. Instead, Parikka and Sampson reveal how the "anomalous objects" of technology are central to its sociotechnical development, and Brunton provides a detailed historiography of the emergence of a specific media form.

Media, Communications, and Cultural Studies Research

In contrast with the studies outlined above, email seems underrepresented in media, communications, and cultural studies. Searches across prominent journals return disproportionately low results for the word "email" in titles, keywords, or abstracts.[59] In some of the results, email did not return in any of those categories. But when I began to sift through the hundreds of mentions of email in full text searches I found many papers were, in fact, substantially concerned with email, the term mentioned 50 times, for example, and featuring extended discussion. To me this is tantamount to being overlooked. I want to ask these journals: *Here is a media form comprehensively covered in your media journal article but you haven't thought it worth recording it in the title, keywords, or even the abstract?*

In addition, email is often used as method of research rather than the object of study. Although articles about such use weren't included in the general tally, the extent to which email underpins research across many disciplines, including interview, diary, or dataset, is another of its hidden stories—and it raises questions of ethics I will later discuss. As a final point of contrast, note that in *New Media and Society*, "email" appears in the titles of six articles, but there are 76 returns for "Facebook" when searching by title.

Crucial email research does exist in the field of media, communications, and cultural studies; I cite many of these works, or draw from them, in this book. So my point is not to admonish my fellow media scholars. But this absence is intriguing. What makes the omission even more compelling is the central focus that "the everyday" plays in cultural theory. It would appear that the ubiquity of email is so total as to have rendered it all but invisible even to those who relish the study of the everyday. In focusing on the mundane and quotidian nature of email, *Email and the Everyday: Stories of Disclosure, Trust, and Digital Labor* brings this study into the foreground. Having contextualized the field, in the following sections I set out methods and outline the chapters.

Research Design and Critical Approach

Everyone has an email story. During informal conversations with colleagues in the preparatory stages of this book, I was struck by the number of embarrassing, hilarious, and regrettable email instances recounted to me. There's the inevitable "reply all" disaster, when an insulting comment made about a coworker intended for a single recipient mistakenly gets sent to the entire office; and the fits of pique or clandestine affection expressed late at night, causing cold sweats in the author the following morning. I've heard the strategies people deploy to cope with ever-increasing volumes of email, and stories from those who refuse to reply to email outside of work hours—not to mention the nonverbal responses I received: the rolling of eyes and shaking of heads from people who think they're at the mercy of this media form.

My book is an attempt to capture and make sense of such sociotechnical, affective practices in a critically systematic, materially situated, and evocative manner. For this reason, I apply both qualitative and quantitative methods. My research for the book draws on two major online surveys conducted with people in the United States (N = 1,031) and in Australia (N = 1,025), as well as interviews with industry experts, email group moderators, artists and creative practitioners, and people for whom email has raised privacy and ethical issues, such as those affected by the Enron dataset release, itself a highly researched email corpus. The US survey ran in 2017 with people aged 18+ who had used email in the last six months and the Australian survey was run in 2015 also with people aged 18+ who had used email in the last six months. Minimum quotas were applied in both surveys across age, gender, and region in order to achieve a balanced sample. Both the US and Australian surveys used a combination of open-ended questions, predetermined yes or no options, and Likert-scale responses to statements. Participants were also invited to contact me to discuss further their personal opinions and experiences. Utilizing a quantitative and qualitative or "mixed method" survey design allowed me to collect different kinds of data.[60] So I posed closed questions to gauge the popularity or prevalence of certain email practices and platforms (most-used email clients or frequency of email checking, for example) and open-ended questions to seek narrative-based reflections about email usage. On this latter point, the questions were designed to elicit stories about people's everyday use of email. I encouraged people to tell me about their experiences of intimacy or emotion through using email: surprising moments perhaps, instances of regret, embarrassment, surprise or pleasure. But I was also aiming to gather insights and descriptions on the routine nature of email, and so I asked questions about the regular uses to which email is put, such as sending documents, arranging meetings, or contacting retailers and service providers. Using this approach, I was able to map both the extraordinary and banal stories of email from ordinary users themselves. In general, the interviews took place via email although in some cases I conducted the discussions face to face.[61] The methods also involved close analysis and detailed summaries of technical reports, white papers, news articles, court papers, and standards literature. Further information about these sources and the various informants are provided during the chapters.

The Everyday

As someone who has struggled fairly unsuccessfully to apply "high" theorists of the everyday such as Mikhail Bakhtin or Henri Lefebvre, my approach leans toward the grounded, situated, or empirical perspectives of this critical terrain to trace the contours of lived experience read through the routines and vicissitudes of email. That is not to reject the necessity for theoretically informed understandings of what it means to study the "everydayness" of culture. It is, after all, feminist theory that has taught us to regard with a keen critical eye the gendered spaces of the everyday, and to insist on the ways in which these must be theoretically *and* empirically negotiated, via, for example, Donna Harraway's work on "situated knowledges."[62] But it *is* to recognize that approaching the everyday analytically presents a number of challenges. For one thing, the everyday often seems resolutely to resist definition: "obvious but elusive," as Fran Martin puts it.[63] The everyday when applied as a descriptive term is ambivalent because it attempts to capture that which we embrace, or at least accept, as pointing to the experiences, feelings, and practices that constitute the major part of our lives. But it may also register a criticism, that which is unexceptional and without merit. As Ben Highmore explains:

> On the one hand it points (without judging) to those most repeated actions, those most travelled journeys, those most inhabited spaces that make up, literally, the day to day.... But with this quantifiable meaning creeps another, never far behind: the everyday as value and quality—everydayness. Here the most travelled journey can become the dead weight of boredom, the most inhabited space a prison, the most repeated action an oppressive routine.[64]

When the everyday is seen through a prism of value and the banal is easily dismissed it becomes fertile analytical territory. As Joe Moran notes, in "cultural studies, the banal is usually turned into something else, made interesting and significant by acts of subaltern resistance or semiotic reinvention."[65] Moran's comment hints at a further paradox of the everyday as it is studied. If we look closely at the ordinary uses of media, we do (by necessity) often remove these objects from their original contexts. By focusing intently on a media object, we risk blunting its everyday nature. As Judy Attfield so astutely notes, "to enclose 'the everyday' in inverted commas changes its meaning ... by plucking it out from the commonplace of the given."[66]

A central challenge thus issued by considering the everyday as an object of critical study is how much to include. Fastidious levels of detail can actually end up fetishizing the practice or media form you seek to analyze and hence undercut its defining quality: that of going unnoticed. Stare at something long enough and it consumes all of your vision. But not registering enough detail tends toward abstraction and, again, loses the spark of recognition that makes the ordinary such a compelling site for analysis. In thinking through how to represent the everyday, scholars have turned to the literature of the French experimental writer Georges Perec. My intention is not to explore his work per se but to draw on some insights from those who have explored Perec's project because of its fascination with detail and inventory, in Highmore's words "acknowledging the world as inexhaustible."[67] For Anthony McCosker and Rowan Wilken, it's crucial to understand the everyday details of our engagements with digital media as rhetorical strategy, to capture "the ordinary intimacies and banalities of life-worlds and media use against the prevailing focus on crisis, protest, or viral flow."[68] Likewise, Perec offers to Highmore methods for documenting the everyday and, again, its attendant paradoxes. In *An Attempt at Exhausting a Place in Paris*, Perec, as the title suggests, attempts to record a visit to Paris. As Highmore notes:

> If exhaustion is the aim, would you need to describe each and every vehicle? Or, is it enough to say that there was a fairly constant stream of foot traffic and road traffic? And, how much should description include? As soon as adjectives are introduced, the flat itemising of objects becomes something closer to interpretation.[69]

Here is what attracts me to a study of email and how it pervades our everyday lives, particularly those days spent at work. Well versed in the office as workplace I am drawn to how these spaces operate in routine and unremarkable ways yet at the same time can provoke moments of passion, sometimes anger, and also pleasure. Instructive in this regard is Melissa

Gregg's study on the impact of media upon office life. Her evocative black and white images of workspaces with their hulking desktops, snaking cords, and dull, flat lighting, bring to life the intimacy of the work environment. She notes the rush of recognition when seeing its representation in Ricky Gervais's highly successful series *The Office*, where those workers depicted "need to believe that collegial pleasures and intimacies can be carved out of even the most hopeless management cultures."[70] I would add a slightly different take on the series, locating it within an aesthetic tradition that tries to represent the ordinary. Here I include the Swedish photographer Lars Tunbjörk, whose work explores the mundane flatness of everyday office life while simultaneously capturing it in beautiful images. His critical eye, the constant oscillation between the banal and the extraordinary, has informed my approach to documenting the everyday. Looking at the everyday contexts of email reveals how, nestled within the sometimes smooth, undifferentiated plane of routine emailing, there are moments of surprise and shock: moments where email communication becomes odd, unfamiliar, and at times perhaps even exotic. As a number of scholars of the everyday have remarked, our critical frameworks need to be robust and imaginative enough to capture a wide range of experiences, social relations, spatial configurations, and technological situations that make up day-to-day life. Alasdair Jones, for example, urges researchers to "pay closer attention to the exceptional as a constituent, imminent, and sociologically vital facet of the everyday."[71] Similarly, in her work on the everyday practices of women's footwear, Victoria Robinson plots out a conceptual framework that recognizes the interconnectedness between everyday moments of drama and banality. As she puts it "the extraordinary is both embedded within and in dialogue with the mundane, rather than having a separate and unmediated existence of its own."[72] It is this sense of "relationality" and interplay, which Robinson so eloquently captures, that I hope to explore throughout the book.

Media and Everyday Life

Scholars have proposed a range of terms to describe the diversified sites of media engagement in everyday life. From Nick Couldry comes the "media manifold," which replaces concepts of "media saturation" to describe the interconnected, complex plethora of media systems, products, and practices that structure digital relations in the contemporary world. Where the notion of saturation falls short is that it doesn't capture the cumulative nature that suggests a more sedimentary structure.[73] Couldry and Andreas Hepp developed the media manifold as a theoretical tool to explain "a large 'universe' of variously connected digital media through which (in various figurations) we actualize social relations."[74] Similarly, Mirca Madianou and Daniel Miller offer the term *polymedia*, "an emerging environment of communicative opportunities that functions as an 'integrated structure' within which each individual medium is defined in relational terms in the context of all other media."[75] Further efforts to describe the totality and particularity of the media landscape include those from Ilana Gershon, who talks about "media ideologies and idioms of practice" to describe the dense media environment and the dynamics by which people are constantly choosing and rejecting a given communication tool.[76] Finally, Highmore retains the idea of media saturation because it gestures toward the hidden or overlooked quality of the everyday itself. As he puts it, "'saturation' could be seen as a cognate term for the everyday: when something reaches saturation point it has bled into the everyday, set up home there, colonized the domestic realm."[77]

I mention these perspectives on media consumption to acknowledge the significant work that has occurred over the last decade about a media environment so inundated with choice. The idea that one can approach "the media" as a monolith has been well and truly debunked. Therefore, to map the terrain, the ambient media of email could be noted. Poised on the edges of the landscape are software applications, hardware infrastructures, and constellations of habit that at first blush seem tangential to the distribution of email yet critically shape its passage. Here we could include the history of dial up, Wi-Fi or broadband networks, electricity grids, servers, printers, and paper. One could also study the relations of practice that are revealed as email functions alongside other message formats. Conventions and modes of conduct begun in email start to bleed into the newer applications of SMS (short message service, or text), Facebook Messenger, and chat programs. Rather than simply replacing email, these other communication systems are used in tandem and so constitute an important backdrop for studying email.[78] But exploring email in a monograph study represents a media choice of its own. While I recognize that email cannot be studied in pristine isolation, I am resolute in efforts to unearth email's significance in everyday life by focusing very squarely on its material and symbolic properties.

Everyday Stories of Disclosure, Trust, and Digital Labor

The title and critical approach for this book reflects themes that emerge from the empirical data collected in the surveys and interviews, and also discovered through the relevant literature. It became apparent to me very early on in the book's design that there exists widespread interest in how others conduct themselves in the less openly discussed aspects of email communication. People seem fascinated to learn about how each other's emails get read, sometimes without the specific authority to do so. During the Australian survey, I was interested to see how willing people were to divulge quite intimate stories in response to the open-ended questions, so I increased the number and scope of these questions in the US survey so as to hear more fully people's "email stories." For example, questions on whether respondents had ever read emails not intended for them provoked an intriguing range of answers. In some cases, these disclosures are accidental, with the humble printer room providing unexpected opportunities to learn about the activities of coworkers or the computer screens that are left unattended proving too tempting a chance to forego. Yet despite access that is inadvertent, participants were still troubled about what that kind of reading practice might mean for issues of trust. Perhaps because of the anonymity that surveys provide, people shared quite compelling stories, the basis of which helped form the various critical narrative threads throughout the book. Here are some illustrative responses about reading other people's emails:

> My partner left email open on the screen, I read out of curiosity, nothing interesting but I felt guilty.

> I guess it isn't technically reading someone's emails, but if people leave their email open in our office in college, we just send them an email telling them they left their email open and to please be aware and then we would log them out.

> I've read my husband's email because his account was open. Didn't breach any trust issues. We have each other's passwords.

> If someone would leave an email up on their screen I'd read it sometimes. I'm nosey.

> It was left open on my screen but minimized. Several hours into my shift when I went to access my email I realized the open screen that had been minimized was not my email but that of the person who left it open before me. I only read what was open and when I realized it wasn't mine I just logged out of the email account to load my own.

At other times however, the reading seems slightly more premediated and perhaps less acceptable than does the glancing moment of unintended access. Quite a few respondents reported having actively read emails, both at home and work, not originally written for their consumption:

> I was in my boyfriend's email looking to see if he was messing with other girls.

> As an IT consultant and member of IT departments, you are entrusted with a vast amount of access to company records and email accounts. Often accidentally and out of no particular interest of my own, I have read through people's emails and gleaned nothing of import but have gained an insight into their day to day existence. At other times, when I have been cynical as to the motives of others in the company viz retaining my services or working to undermine me, I have abused my administrative powers to access their email in an attempt to find out more about what is planned for me. This is often successful, but occasionally without merit and always deeply unethical.

> I often read my wife's emails without her knowledge.

> I read my mother's email when [my] parents were going through a divorce when I was a child.

> I've started a new job and have access to my predecessor's inbox. She used her work email address for a lot of personal stuff.

> When I was in a temping role for a week and it was so boring because there was virtually no work to do, I looked at the personal emails of the girl I was replacing, got bored after half an hour. This was pre-Facebook/other social media.

Such responses represent rich source material for tracing a range of email experiences in daily life. For example, some of these extracts point to the very human aspect of workplace monitoring beyond the sensational headlines we encounter about lawsuits, email breaches, and hacks (although some of these will be discussed). Surveillance can often happen informally at work through the flows of bureaucratic media. It is also worth noting with amusement how the last respondent highlights the almost acceptable and certainly longstanding tradition of workplace procrastination, clearly in place before we took to our Twitter accounts to avoid a daunting task. More generally, the participant views and experiences quoted above show the important role of disclosure, trust, and digital labor in our everyday encounters with email.

Book Structure and Chapter Outline

I've organized this book in three parts, each offering a different focus for the everyday contexts of email use. One of my central arguments holds that the ubiquity of email has obscured its study. We seem to know so much about other fields of media engagement through the "worldmaps" of social media formation that plot how these players are connected, their business models and algorithms, or how their users access the services. Similarly, important critical work has been carried out across the media industries to glean their political economy frameworks. There is not yet a comparable model for email. So in part I, " Histories and Landscapes," I lay out the technical and historical foundations from which email grew and map its current social and industrial architectures.

In chapter 1, "The Origins of Email and Its Development," I look at some of the key points in the historical trajectory of email including the early exchange of messages across the Compatible Time Sharing System (CTSS) known as MAIL, designed by Tom Van Vleck and Noel Morris; Ray Tomlinson's work and his choice of the now iconic @ in the address; and the growth of email software or "clients," including John Vittal and the MSG program. But as we'll learn, definitions of email are not straightforward. In the late 1970s, the United States Post Office (USPS) attempted to redefine their business model so they could take advantage of the emerging system of electronic mail. In fact, the dialogue between the postal service and email has been a feature of the communications landscape for more than 40 years. Internet governance plays a key part in how emerging standards and conventions are developed, so I also explore the cultural role of the RFCs (Request for Comments) and the US-based Internet Engineering Task Force. I then look at the memo template of email, those "to" and "from" fields still highly visible to us today. I am interested in the fundamental role the header format plays as a mode of organizational media that shapes our institutional life.

While in the first chapter I may touch upon the business of historiography—what is selected as significant and what is omitted—in the following chapter I bring this squarely into focus. Chapter 2, "'Inventing Email' and Doing Media History," concerns a fascinating moment in email history with the controversial claim by V.A. Shiva Ayyadurai that he, not Tomlinson, invented email. As a media researcher, I was quite conflicted about whether to include Ayyadurai's assertions. For many in the field of internet studies, this event warrants nothing more than a footnote in history. To devote serious attention muddies the waters of historical accuracy. But once again, as a media researcher it is difficult to ignore the complex strands that weave together legacy media, celebrity cultures, and law, so I hope to have painted a suitably compelling picture to explain its inclusion.

It also represents a case in point about the dance between the extraordinary and the banal that one encounters in the study of email.

In chapter 3, "The Email Industry," I move from histories to landscapes. In order to understand email properly we need to look at how it organizes itself as a field. Here I am talking about the email service provider industry, which is constituted by a complex range of commercial and open source organizations, applications, and architectures. To convey a sense of the interconnecting parts of the industry, I spoke with a number of email providers, in particular Rob Mueller, one of the founders of the Australian company Fastmail, and Nicola Nye, its chief of staff. Underpinning the success and failures of many of these organizations are email analytics. We are now familiar with how algorithms fundamentally shape our daily activities online but perhaps give little thought to the function of *opens*. This metric, along with *bounces*, *forwards*, and *prints*, is key to the email economy and can provide a snapshot of everyday media use. However, just as with the opaque analytics of social media, email measurements are often hard to parse. Sourcing independent, noncommercial, or objective figures on worldwide email use is quite challenging. A software company whose primary business is providing security solutions is unlikely to discover evidence of widespread best practice email habits. Businesses offering alternative tools to workplace email are probably going to find increasing levels of frustration with existing modes. Such realizations motivated me to run to the two international surveys devoted to exploring the scale of email consumption and to glean insights into people's habits and attitudes around email practice.

In the second part of the book, "Affect and Labor," I pick up on an earlier point concerning the memo and how it shapes institutional uses of email. For many of us, work is the primary site of email communication, so in chapter 4, "Bureaucratic Intensity and Email in the Workplace," I look at its impact in professional and organizational settings. Despite inroads made by alternative collaborative workflow tools like Slack, email still dominates the workplace. Indeed, the exact same arguments currently advanced for the superiority of Slack, its flattening of office hierarchy, were made about email in the 1980s. Drawing from the surveys and interviews, I discuss how people engage with email at work and examine their perceptions about the conventions of organizational hierarchy, especially with the use of emoji, as well as their practices of using email for personal reasons at work. Here

I make the point that the role of affect in our everyday office encounters is underestimated. To tease out what this impact might be I use the term "bureaucratic intensity" to capture the rhythms of workplace email, its flashes of anger, its moments of pleasure, and its routine nature. While bureaucracy might seem an odd topic for the discipline of media and communications to investigate, I argue it names a particular set of conditions that operate in the workplace. Together with spreadsheets, forms, vision statements, and minutes, email underpins the media ecology of bureaucratic life, regulating practice and identity in significant ways. The literature on bureaucracy as an institutional system has tended to overlook the role of affect in structuring the workplace. In response I turn to feminist accounts of workplace gender and power, revealing how kindness, support, and patience are routinely ignored in workplaces yet whose benefits actually shore up more formal economic and social structures.

Labor and affect are again my focus in chapter 5, "Moderation and Governance in Email Discussion Forums," where I investigate one of the most enduring uses of email: that of the list or group. Although social media governance has occupied the research agenda of communication scholars, its roots in online discussion forums have been largely ignored. Not concerned with a sweep of history, however, I aim in this chapter to bring to life the contemporary vibrancy of email groups for building social networks, resource sharing, and professional development. Email lists often gain strength through their narrative capacity. This seems particularly true for those who subscribe to health-related groups, where telling your own story or identifying with someone else's provides a powerful support tool. In mapping the function of email discussion groups, I explore the role of moderators as they perform important work mediating between the software constraints of a mail distribution system and the voice of a particular community. I discovered that moderators were often reluctant to categorize their duties as "work," and this hesitation is read alongside the theories of "digital exploitation." Three themes emerged from talking to the moderators: affective management, perceptions of labor, and the future of email lists.

In the final section of the book, "Archives and Publics," I explore how email occupies and constructs the public sphere, and I chart its path to the archives. In chapter 6, "The Enron Database and Hillary Clinton's Emails," I recount what happens when private or personal emails become public by exploring two case studies where emails written in unexceptional circumstances suddenly get thrown into the limelight. Email hacks and leaks inhabit our popular imaginary about the fate of correspondence in a post-Snowden world, but what about the people who get caught up in public release of these vast datasets?

The first case study considers the "Enron corpus," the largest publicly available test bed for email research. In 2003, internal emails from the energy giant Enron were released into the public domain by the US Federal Energy Regulatory Commission (FERC) during its investigation into the company's fraudulent activities, which had resulted in criminal findings against a number of its top executives. Since then it has become a key source of data-mining projects for the research community, attracting diverse studies across management theory, organizational behavior, fraud detection, counterterrorism, social networking sentiment analysis, gendered language usage, and mobile interface design.

But little has been said about the people who for many years have had their emails intensely investigated, printed, parsed, and displayed. I spoke with a number of these people about their experiences of having emails that were written with the expectation of privacy released to the public domain. Here is where the dance between the extraordinary and the everyday becomes particularly salient. When viewed from the outside this case study may seem exotic: a tale of extreme company malfeasance, drama, and corruption. Seen from the employees' point of view, however, their email experience was indeed very "matter of fact"—until their work emails became the subject of public scrutiny. Moreover, the drama of this act was then itself made more ordinary by the circulation of this corpus through the research communities on the internet in their everyday work situations.

The second case study concerns the Hillary Clinton emails, a dataset that became public in 2015 when it was revealed Clinton had been using a private, unsecured email server for official documents exchanged in her role as US secretary of state. Rather than follow an overtly political trail of interpretation, I read this as an example of the everyday uses of email and the emerging prevalence of shadow IT, familiar to us from applications like Dropbox, WhatsApp, or Gmail, in workplace communication. From looking at the perils of metadata here, in the final chapter I consider the imaginative uses to which it has been put.

In chapter 7, "The Art of Email," I explore the creative possibilities of email identifying its trajectory from the epistolary novels of the eighteenth century to their modern counterparts in the "email novel," along the way touching upon the vernacular expression of ASCII art and "typewriter art." Here I look at four projects spanning the last 40 years to show how artists have responded to the aesthetic opportunities of email: Carl Steadman, *Two Solitudes, an e-mail romance* (1994); Miranda July, *We Think Alone* (2013); Daniel Smilkov, Deepak Jagdish, and César Hidalgo, *Immersion: a people-centric view of your email life,* (2013); and Brian Fuata, *All Titles: The Email and SMS Text Performances* (2010–2016). For each of these case studies I spoke with the artists to ask questions about the project inspiration and their objectives in exploring the aesthetic possibilities of email. Animating many of the artworks under consideration is the capacity of email to tell stories. In the email novel the narrative is driven, as the epistolary novel before it, by the writing, sending, and delivery of email. But as I want to explore throughout the book, this storytelling power is not limited to fiction. In the artworks considered, as in many of the instances touched upon in the book more generally, stories told about and through email are one of the key ways by which it becomes stitched into the everyday.

I Histories and Landscapes

1 The Origins of Email and Its Development

Reading email in the 1990s is accompanied by the sound of dial-up. It is opened on a stationary monitor and the screen measures about 14 inches. Enabling access to this email is a hard drive with 1,000 megabytes capacity encased in the aptly named "tower" that is tethered to the wall through power cables. Unless you print this email, it is nearly impossible to carry with you. If you do print the email it will almost certainly appear in black and white ink and emerge from a machine weighing more than twice as much as a printer today.[1] While screen estate and mobility have changed dramatically during the last two decades, the point to make here is not so much about a desire to capture the pace of technological development as it is to focus attention on the material and discursive assemblages of email and the circuits through which it travels. How does the memo template make meaning and produce sociality? What are the internet protocols, interface design decisions, and cultures of production lying behind those familiar fields "to" and "from"?

Since my central claim in this book is that email's ubiquity has obscured its study, in this chapter I attempt to make visible its materiality, its look and feel, by exploring the development of email through a number of key moments and sites. Beginning in the 1960s where electronic messages were sent within a single system to the early emergence of networked mail in the 1970s, the story of email also intersects with that of the United States Post Office with bids to expand their business into the exchange of electronic message. The history of email also includes looking at the distribution system and the standards known as SMTP (Simple Mail Transfer Protocol), POP (Post Office Protocol) and IMAP (Internet Message Access Protocol). Perhaps these initialisms and acronyms themselves are not familiar but they form the basis of all our email communications. Similarly, if you encounter the term MIME (Multipurpose Internet Mail Extensions) you may not initially recognize its significance. But you are almost certainly accustomed to the function of "attachments." For respondents in my US survey, sending photos, music, or documents was one of the most popular uses for email. Protocols are the result of complex social and technical dynamics, so in this chapter I also consider the central role played by the long-standing Request for Comments series (RFC) and the US-based Internet Engineering Task Force in creating these developing standards. What emerges from these socio-material settings is the memo template, an email design format that, I argue, has become significant in shaping the daily domains of life.

Early Days

As briefly touched upon and generally accepted by the internet community, Ray Tomlinson created the first network mail program in 1971 while he was an engineer at the US computer science research firm Bolt Beranek and Newman Inc. (BBN) which later became the defense contractor Raytheon BBN Technologies.[2] BBN was instrumental in building the first nodes of the ARPANET, precursor to today's internet. In a much-quoted passage, Tomlinson explained his contribution:

> The first message was sent between two machines that were literally side by side. The only physical connection they had (aside from the floor they sat on) was through the ARPANET. I sent a number of test messages to myself from one machine to the other. The test messages were entirely forgettable and I have, therefore, forgotten them. Most likely the first message was QWERTYUIOP or something similar. When I was satisfied that the program seemed to work, I sent a message to the rest of my group explaining how to send messages over the network. The first use of network email announced its own existence.[3]

Tomlinson also chose the @ sign to distinguish between the username and the name of the host computer. It was a design decision he would later come to question. Originally it was picked for its neutrality because it did not appear in a person's name and carried no executable effect on TENEX computers. Unfortunately, it turned out that for users of the Multics system this @ sign read as a "line-erase" command. Once the software encountered the sign it would delete all associated characters in the address. This mattered because at the time Multics and TENEX operating systems dominated communications infrastructure, and a situation where the two were incompatible was certainly not ideal. Discussing this aspect Tomlinson later admitted that he "might have been friendlier to the Multics folks" in choosing the @ sign.[4] Rivalry between the BBN-based TENEX system and others like Multics actually produced significant email advances as these engineers debated what protocols the nascent internet developers should adopt. Tomlinson's significance has been extensively acknowledged. He was inducted into the Internet Society's Hall of Fame in 2012 for inventing networked email and more broadly for his contributions to designing early versions of SMTP, the standard for email transmission in use today.[5]

When trying to understand the history of email, it's important to note the existence of electronic message exchange prior to Tomlinson's contribution. During the 1960s it was possible for users of the same computer to communicate with one another. At MIT, the Compatible Time Sharing System (CTSS) ran a program called MAIL, written by Tom Van Vleck and Noel Morris, which provided a mailbox where people could send and receive messages.[6] In 1965, newly appointed as research staff in the Political Science Department, the two read a proposal written by Louis Pouzin, Glenda Schroeder, and Pat Crisman for a message facility to be added to the CTSS. Discovering that no one had the time to actually write the program, Van Vleck and Morris set about putting it into action. Refining the original design, their software now included a specific command for reading email whereas previously the print command had been used. Before their program was implemented, people could communicate by leaving messages in "common file" directories or folders. This offered little security—anyone logged on could read the messages that had been left—and it would only work for users who had access to the same directory. Instead the MAIL program now provided the capacity for anyone logged in across the entire CTSS to communicate, and it also supplied the functionality of a new private mode. The program was soon enthusiastically adopted across the network, some 1,000 people, who would discuss research-focused topics but also began using the infrastructure for personal and social purposes.[7]

Despite its obvious popularity however, the CTSS MAIL program enabled communication only within the same computer system. Tomlinson's contribution was unique because he made it possible for mail to be exchanged between different computers. As Craig Partridge explains, Tomlinson achieved this by building on the existing SNDMSG software, a local host-specific email program in use on the TENEX operating system, to incorporate features from the file transfer system called CPYnet, a program that had also been written by Tomlinson. Eventually CYPnet was replaced by what would be the first generation of standardized File Transfer Protocols of today.[8]

The mid-1970s saw a number of significant email advances, with Stephen Lukasik, ARPA director from 1971 to 1975 and one of email's "great advocates," being the driving force behind many of these innovations.[9] Although email was gaining traction across the network, its use was still limited mainly to computer scientists. Lukasik sparked its sudden widespread uptake by encouraging his staff to communicate with him through that means. Indeed, as Janet Abbate argues, it was this very "access to power" that email enabled; budgets and funding seemed to come a little easier to those at ARPA who could attract Lukasik's attention. Those without email felt disadvantaged. But while he extoled its virtues, at times email was unwieldy for Lukasik. Restrained by having to read through all messages in order, Lukasik complained to Larry Roberts, himself an early email adopter in his role as ARPANET pioneer, to design a replacement for the current unwieldy READ-MAIL. In response Roberts wrote the mail manager program known as RD, which now "made it possible to list the messages in the mailbox, to pick which message to read next, and to print individual messages."[10] In turn this was enhanced by Barry Wessler's software dubbed the New RD (NRD), which added features to file, retrieve, and delete messages. Partridge calls RD and NRD the "first mailbox management tools and the first true user agents."[11]

User agents, or email clients, began rapidly to develop from this point with John Vittal a leading figure. Expanding the existing BANNARD program that had been developed by Marty Yonke, itself elaborated from Robert's RD, Vittal had a keen eye for whether email was actually meeting people's needs. Listening to feedback, his MSG program transformed how emails were managed, and he introduced the ANSWER command recognizable today as Reply.[12] What this now made possible was the automation of a hitherto laborious process. No longer did email users need to manually type out the recipient address and subject line. Also supported by this new feature was whether to reply to a single author or to use the now notorious "reply all" function. Partridge suggests the ANSWER command could have sharply increased traffic across the network.[13]

Vittal's program was so groundbreaking that he has sometimes been hailed as the true inventor of email.[14] Described as the "hacker's hacker" by Katie Hafner and Matthew Lyon, Vittal's significance is in part due to

Figure 1.1
Email using the DEC VT52 computer terminal, circa 1978.
Source: Photo by Dennis Coady. Image Courtesy of the Computer History Museum.

the approach he took to software development. Originally conceived in 1974 while he was working at University of Southern California's Information Sciences Institute, the implementation of MSG was not institutionally funded. This meant it was Vittal's own project, built in his spare time. So when he moved to BBN in 1976 his work on MSG continued in this informal, ad hoc manner.[15] Notwithstanding its unofficial, nonproprietary status, or perhaps because of it, MSG spread quickly through the scientific research community partly because it was so easy to access. As Elizabeth Feinler (the former director of Stanford's Network Information Center) notes, MSG was widely copied across the internet: "it was a freebie available to anyone who wanted to install it." There were also reports that senior members of the Pentagon had adopted MSG, and at BBN in the 1990s it was still in use.[16] In this way, MSG is a notable precursor to contemporary forms of "shadow IT," those systems and programs deployed by institutions that nevertheless do not endorse (and sometimes actively denounce) their rollout. Examples here are Gmail and Dropbox, which I explore further in chapter 4 in terms of work and bureaucratic intensity.

Meanwhile let's return to the significance of MSG. Along with the ANSWER command, chief among its innovations was the "forward" functionality and a command that enabled messages to be moved, saved, or deleted. The MSG program offered a prototype for the design of the Pentagon-funded email program HERMES, and it also inspired the subsequent standardization and expansion of the email header field.[17]

If Vittal's MSG program saw widespread success, Douglas Engelbart's email program was rather less fortunate. In the late 1960s and early 1970s a MAIL feature was included in the oN-Line System (NLS), an information and resource sharing computer network Engelbart established at the Stanford Research Institute's Augmentation Research Center (ARC). Unlike the rapid uptake of the prominent tools of hypertext or the mouse, the MAIL application on NLS found it impossible to gain traction across the wider ARPANET user groups. While popular within ARC, MAIL could not compete with the programs favored by the TENEX applications SNDMSG and RD.[18] Abbate suggests this was because NLS was a difficult and unwieldy program to master.[19] For Thierry Bardini the unpopularity of MAIL was bundled up with wider administrative, funding, and personnel issues at play at ARC that would eventually see the center dissolved.[20]

Counter Claims

At one stage or another, most historical accounts will face some form of contestation or elaboration. One such instance occurred when Mike Padlipsky, an early member of the ARPANET design group, published his essay "And They Argued All Night ..." in 1998. With this article Padlipsky wanted to clear up the confusion. It is time, he wrote, to "dispute various self-serving claims of Fathers of this and Inventors of that." For Padlipsky a number of factors came into play, including the term "email" itself, and the significance of the existing File Transfer Protocol. As he explained:

> I don't believe Ray Tomlinson invented "e-mail." And not because of the quibble that we called it netmail originally.... Nor because of the semi-quibble that "mail" had been around intra-Host on several of the Host operating systems.... No, it's because I have a completely clear memory that Ray wasn't even at the FTP meeting where we decided to add mail to the protocol. Granted, one of the BBN guys mentioned that he'd done a TENEX to TENEX mail hack already, and that he'd used the "@" between id and host name for the addresses, but that was after somebody had said, "Hey, why don't we send mail via FTP?" *And for all I know, that somebody might have been me.*[21]

Others in the software engineering sector have taken up Padlipsky's "quibble." For some, the addition of the qualifier "electronic" was redundant. The term "mail" already described a system for transporting messages: Why, then, add a reference to the specific carrier? As Van Vleck explained in a 2013 *Wired* piece, "I don't really like to use the term 'e-mail' or 'email.' I usually just call it 'mail.' The use of electrons for mail may someday become quaint, replaced by photons or quarks; should we prepare to speak of 'p-mail' or 'q-mail'"?[22] Van Vleck has often asserted the role he played in the development of email. Interviewed in 2011 for a *New York Times* story about computer history he wondered whether the difference between the mainframe messages and the subsequent invention of email was really that significant. Asked about Tomlinson's contribution, Van Vleck responded:

> Everybody said, "Well, gee, Ray did this great thing. He sent electronic mail from one computer to another." But how did they know it was electronic mail? The TENEX machines that he used all had internal electronic mail just like CTSS had.... TENEX had its own, internal mail command.[23]

In the interview he conceded that what Tomlinson did was to "figure out how to break out of one box and get it to another box, using the new ARPANET."

But isn't the transfer of typed messages between geographically dispersed agents the foundational feature of email? Van Vleck then elaborated:

> My thought is I walk into my office. I sit down at my terminal. I type a message and say, "Send." And someone else in some other room sits down at his terminal and reads it. Who cares how many computers there are? Who cares whether the computers are owned by the same agency or different agencies? That was another argument. It isn't e-mail unless it is between separate administrative domains.[24]

If one holds fast to a view of technological history that identifies a single point of origin and utilizes a language of "firsts," "pioneers," "visionaries," and "inventions," then statements like those quoted above can be couched in narrative terms of "counter claims." The original facts from a certain individual are weighed against subsequent claims, all with the objective of proving once and for all who invented said technology. Scholars trained in science and technology studies have critiqued such a view pointing out how innovation and development is frequently the result of collective efforts rather than individualist heroic triumphs; that attention on the individual eclipses the wider cultural, political, or economic contexts from which a new technology might emerge. I will be looking at this further in the next chapter when I discuss email history more broadly by exploring an instance where media, technology, celebrity, politics, and law form a complex set of conditions through which to view the question of "invention."

ECOM and the United States Postal Service

The tussle over email definitions was not limited to the world of programmers and engineers. During the late 1970s the United States Post Office (USPS) weighed in on the debate when it tried to redefine its core business to include email. At the time, the USPS was proposing a new service called Electronic Computer-Originated Mail (ECOM or E-Com) that would deliver paper documents generated from messages sent electronically. In its application to the Postal Rate Commission, the USPS described a system where "messages would be received at the post office close to the addressee on high-speed printers, and would then be automatically … folded, inserted and sealed in envelopes." Subsequently, the "messages would be entered in the mailstream and processed for delivery with first class mail."[25]

Not surprisingly the proposal was met with opposition from those in the private sector who feared this paved the way for the USPS to become a telecommunications service provider. AT&T, for instance, petitioned strenuously against the initiative. So successful were the protestations from commercial lobby groups to the Postal Rate Commission that ECOM was forced to substantially revise its final proposal. Originally ECOM had wanted to create its own electronic network for transmission, known as third-generation mail, but now it had to outsource this aspect to those telecommunication providers.[26] In its review, the Committee on US Postal Service Planning for Electronic Mail Service Systems explained that the provision of an "electronic end-to-end delivery system … is deemed to depart too much from the basic character of the Postal Service and to approach privately supplied communications services too closely."[27] Reactions from the computer engineering community were mixed. Some had already been advancing arguments that the Post Office should increase its remit to include electronic mail. In their 1977 paper "The Convergence of Computing and Telecommunications Systems," David Farber and Paul Baran urged network users to embrace the inevitability of automation. Following the modernization of financial and medical records, they cautioned, "electronic transmission of much of the current First Class Postal mail similarly will follow."[28] But others bristled at the suggestion that the USPS could determine the future of electronic communication. At the 1980 Electronic Mail and Message Conference held in Washington, DC, Ted Nelson was in attendance and reported that the USPS representatives behaved as if someone were "taking away what rightfully belonged to the Post Office."[29]

Nelson was onto something. Part of the implementation process for its new service involved the USPS seeking to broaden the scope of its mail carriage to cover data processing as well as electro-mechanical and electronic processing. Moreover, it wanted to change the statutory definition of "letter" to include the phrases "floppy disc" and "orientations of magnetic particles."[30] The reason for this move was simple. Currently the USPS held a limited monopoly over letter transport so that competing players could not enter this segment of the market, but parcel and telegram mail categories could. In order to bring future electronic mail business within its reach, the USPS needed to expand the definition of letter or, to put it another way, to define email as letter writing. Unfortunately for the post office, the US Justice Department Antitrust Division did not endorse this USPS legislative reform, calling its expanding definitions of letter traffic "exotic."[31] ECOM, however, continued with its plans and launched the program in 1982. But the initiative was short-lived, with estimates for its revenue losses exceeding $40 million.[32]

Intelpost was a similar product in operation during this period. By describing it as a "digital facsimile network between US and foreign post offices for international electronic mail," the USPS was again attempting to bridge the gap between hardcopy mail and the telecommunications market. The initiative involved delivering messages to an Intelpost office where they would be "scanned by a facsimile reader and transmitted via satellite to their destinations." Once received, the messages would be "reproduced on paper by facsimile printers, inserted into special envelopes, and delivered by the post office in the participating countries."[33] These countries included the United Kingdom, Sweden, Germany, France, and Australia. Interestingly, Tomlinson reported that he had been quite involved in the discussions about the design and implementation of Intelpost technology during the late 1970s and early 1980s.[34]

The Social and Technical Emergence of Email Protocols

The early 1980s was a busy period for email development. While the USPS was attempting to expand into new electronic markets, email itself was developing apace through advances made in its distribution systems and software. This section briefly explains the underlying technical protocols for email and the Request for Comments (RFC) system through which various design decisions are made. As we will see, the RFC series helps fashion contemporary email practice and yet its sheer density of information and the frequent opacity of its administrative processes have sometimes actually threatened to forestall development. Likewise, the memo template, while easily dismissed as a banal media form, has a significant impact on contemporary email design and its social uses.

Email transmission functions through the Simple Mail Transfer Protocol (SMTP). This technical internet standard determines the delivery of email whereas receipt and download occurs via the Post Office Protocol (POP) and the Internet Message Access Protocol (IMAP). POP and IMAP differ in the degree to which they allow a user to interact with a remote server to manage mail. In general IMAP permits email to remain archived on the host mailbox while POP enables email clients to delete messages once they have been downloaded. During the costly days of dial-up, POP was a significant innovation because users did not have to remain connected to their ISP in order to read email but could download their messages for later use.[35] These communication standards developed through the RFC (Request for Comments) system set up by Steve Crocker at UCLA in 1969 in order to document and discuss the emerging conventions of the ARPANET, the precursor to today's internet.[36] The original SMTP was outlined in RFC 821 by Jon Postel in 1982 and has been updated on a number of occasions to improve issues of security and legibility on mobile devices, mainly through RFCs 2821 and 5321.[37]

Although SMTP is the most widely used transport mechanism for email it has faced competition from the parallel standard called X.400. First developed in 1980s by the International Organization for Standardization (ISO), X.400 is sometimes thought superior to SMTP in terms of better security and more sophisticated message-handling features. Equally, it is often characterized as bloated, its address syntax derided because of its unnecessary complexity. Debates about the respective utility of these two standards are themselves embedded within wider discussions concerning internet governance and architecture illustrated by the rivalry between the network conventions of Transmission Control Protocol/Internet Protocol (TCP/IP) and Open Systems Interconnection (OSI). Since X.400 emerged within the suite of products developed by the OSI initiative, detractors and defenders of its capacities tend to view it through the prism of what Andrew Russell calls the "Internet-OSI standards wars."[38] This refers to a period of internet history between the mid-1980s and early 1990s when the OSI protocol looked set to replace TCP/IP as the vendor-neutral universal standard connecting the growing constellation of computer networks.

At the time, TCP/IP was widely used but there was general agreement it could be improved to ensure global interoperability. Initially, the OSI had support from the US government together with major stakeholders worldwide including engineers, telecommunication suppliers, and computer companies. But it wasn't to be. By the early 1990s the internet's own TCP/IP suite became the de facto language of connectivity. This success did not happen overnight. While exacting deliberation occurred at the theoretical-standards level to implement OSI, in everyday practice TCP/IP was already hard at work. For many, this was proof that OSI only existed as an abstract, impractical ideal, summed up in the proclamation of TCP/IP advocate Einar Stefferud that "OSI is a beautiful dream, and TCP/IP is living it!"[39] This rather anodyne phrase belies the intensity of feeling and the institutional and commercial exigencies at play. For many, the failure of OSI demonstrated the very worst of "rule by committee." The ISO was cumbersome, an unwieldy model of European decision-making and "the authoritarianism of a sclerotic bureaucracy," as Juan D. Rogers and Gordon Kingsley wryly noted.[40] Others have pointed to the obvious technical, real world success of the network protocols already in action. For Russell, such tensions are illustrative of the social and political dimensions of standards-setting culture. As he puts it, standardization is a "contested process whose success depends upon the obfuscation of its founding conflicts and contingencies. Successful standards, if they are noticed at all, simply appear as authoritative, objective, uncontroversial and natural."[41] What the battle between TCP/IP and OSI highlights for Andrew Russell is a problem with the prevailing internet narrative that claims its founding architectures emerge from democratic, open settings. Instead, he has argued that the triumph of TCP/IP illustrates the operation of a profoundly autocratic and centralized system of decision-making:

> In a strange plot twist, the autocratic Internet emerged as a symbol of open systems and an exemplar of a new style of open-systems standardization.... The Internet soon became known as an unqualified success story, and its advocates quietly recast Internet history as a story of open and democratised innovation—ironic as its sponsor, the American Department of Defense, was the quintessential Cold War closed world institution, and the leaders of the Internet community repeatedly rejected basic formalities of democracy such as membership and voting rights.[42]

The passage of time has not soothed the antagonism of this debate. In 2013, when Russell published an online version of his research, the comments section lit up. Despite 20 years having elapsed, discussion raged about the merits and downfalls of the two networking standards and the sociopolitical institutions from which they had emerged. Still very much alive is an open disdain for bureaucratic forms of internet governance, with one poster guffawing about French committee delegates and their predilection for outlandish menus over actual technical work in the field. Commenters also drew attention to the open form of the RFC system against the costly membership fees demanded by the ISO.[43]

Returning to the two email conventions of SMTP and X.400, we see how social and political desires often drive technical implementation. If the OSI fracas had not occurred, the latter protocol could well be more widespread today. Craig Partridge notes that although X.400 had many design advantages over SMTP, for example the inclusion of third-party content, the computer engineering community "tarred X.400 as a component of the unpopular OSI protocol suite."[44] Comparing the SMTP and X.400 debates to other historical standard wars like Betamax and VHS, Kai Jakobs argues that the problem wasn't precisely that these specifications got caught up with the broader TCP/IP success. Rather, the standards system had not been responsive enough to rapid technological development. When its first implementation was released to the market, the associated problems could not be addressed in time, so the reputational damage had already been done.[45]

Attachments

Initially, email communication was text-based only. It expanded to include non-ASCII characters—in other words, it moved beyond the American Standard Code for Information Exchange, which devised a numeric system to represent all 128 English characters. Email communication also expanded to handle HTML, audio, video, or image file attachments through the development of Multipurpose Internet Mail Extensions (MIME), which was first proposed in a series of RFCs during the early 1990s by Nathaniel Borenstein and Ned Freed. If these authors are said to have invented the attachment, this wasn't their original objective. Instead, Borenstein envisaged the inclusion of an object into the actual body of the email itself rather than have it appended using the ubiquitous paper clip icon.[46] The consequence of this design decision is that different email clients and devices manage attachments in multiple and conflicting ways; a user may embed content that is incompatible with the destination software or might be unable to attach a file from a particular device or application. Without leveling all the blame at the invention of MIME (which offers myriad internet functionalities beyond email), we can undeniably say that attachments are often the source of intense irritation. From eating up our bandwidth and threatening our hard drives with malware to issues of legibility and screen size, attachments are central to how email produces affect in everyday life. Commercial terms of use and email guides brim with stern warnings to users about the perils of attachments sent in haste or carelessly opened. Typical of such guidelines is the familiar exhortation to open only those attachments sent from a trusted source in an effort to reduce the risk of spam. At the institutional layer, limits are often imposed on the types of files sent where programs or executable attachments are filtered and deleted automatically before reaching the user. Organizations, service providers, and email software clients can also limit the size of email attachments; 25 MB currently operates as an industry standard.

Increasingly, third-party applications like Dropbox are used to resolve some of the drawbacks and software conflict issues encountered with email attachments. Instead of sending a large file, a link is included that directs to a cloud-based storage system. Since writing the RFCs for email attachments Borenstein has become the chief scientist for Mimecast. Named with an obvious branding nod to the MIME protocol, this cloud-based email service provider offers outsourced archiving and security applications for commercial organizations.

Reading the RFCs

At the time of writing, there are more than 6,000 RFC documents archived by the Internet Engineering Task Force; the system remains a significant social and technical driver for internet governance and operation. Originally, during the early 1960s, the Network Working Group (NWG) managed the distribution and storage of RFCs. NWG members established, via the RFC titled "Documentation Conventions," the egalitarian and collaborative ethos that has since characterized its operation as one of "rough consensus."[47] As Steve Crocker stipulated: "Notes may be produced at any site by anybody and included in this series," with the content of the notes to be expansive and inclusive, comprising "philosophical positions without examples or other specifics, specific suggestions or implementation techniques without introductory or background explication, and explicit questions without any attempted answers"; then he added that "the minimum length for a NWG note is one sentence."[48] Aside from defining the core protocols of email communication, the RFC series has been a central site for developing internet technical standards including File Transfer Protocol (FTP) outlined in RFC 765 and Hypertext Transfer Protocol (HTTP) described in RFC 2068.[49]

The RFC series publishes documents organized according to different categories of internet governance including *standards, best current practices, historic, informational,* and *experimental*. Each publication is formatted as a memo-style template and provides the RFC sequence number, author name(s) and institutional affiliation, category of submission, date, and abstract. To qualify as an internet protocol a paper cycles through a peer-reviewed process of *proposed standard, draft standard,* and finally *standard*. Mailing lists of the various stakeholders play a central role in peer review. The RFC editor manages the process; Postel held that position until 1998 when his collaborator Joyce Reynolds took over the job. The editor's role has evolved over the years supported by a network of funding bodies, associated committees, and personnel. At the time of writing, the RFC series editor is Heather Flanagan, assisted by an Advisory Group and Oversight Committee with interface design and search functionalities provided by the business software company Association Management Solutions.

Liaison between industry standards bodies and the RFC community has sometimes been prickly. Where the RFC publications are affiliated with the Internet Architecture Board, the Internet Engineering Task Force, and the Internet Society, tensions often emerge about the editorial procedures, particularly those involving digitization policies of the archive and a move from ASCII templates to HTML.[50] Although the systems governing RFC submissions have formalized over time, contributions remain an open process across the internet, producing a vibrant discursive public of software engineers, app developers, policy makers, academics, legal practitioners, advocacy groups, and the commercial sector.

A fundamental convention of the RFC series is that notes are archived in their original form. Amendments are implemented on subsequent memos which include the term "update" and "updated by" in order to indicate the chronology. Recognizing its potential, researchers have analyzed the RFC corpus as a source of internet protocol history within legal, economic, political, and social frameworks. Sandra Braman examined the entire dataset for the 40-year period from 1969 to 2009 using discourse analysis to show how these technical discussions have made an impact on the policy and legislative design of the internet. Whereas literature concerning the legal response *to* technological development is vast, her research reveals the process by which technical design has an effect *on* law, bringing the two fields into dialogue and shedding light on what she calls their "interpenetration." As Braman explains of the RFC system:

> The technical decisions that resulted had law-like effects in the sense that they constrained or enabled the ways in which users can communicate and can access and use information over the Internet, whether or not such decisions supported or subverted legal decision-making, and whether or not legal decision-makers understood the societal implications of the technical decisions that were being made.[51]

Similarly, Laura DeNardis includes the RFC system as a constitutive site for "protocol politics" where the design of internet architecture shapes questions of security, privacy, and access to knowledge, technical standards whose functions and languages are often difficult to grasp yet far reaching in their geopolitical scope.[52] With these insights DeNardis joins scholars working in the field of software studies focusing on how protocol provides the logic for new forms of power in contemporary culture; Alexander Galloway calls the RFC series the "primary source materials" for such analyses.[53]

Galloway's work has been key to developing a critical taste for the materialities and social functions of computer protocol, defined as a "distributed management system that allows control to exist within a heterogeneous material milieu."[54] For him, scholars like Lawrence Lessig have got it wrong when they imagine total internet freedom was once possible and now lament it has been supplanted by regimes of commercial intellectual property power. Instead, Galloway says, control has been there "from day one" because "control is endemic to all distributed networks that are governed by protocol."[55] Galloway is no fan of bureaucracy. Many praise the Internet Engineering Task Force (IETF) for its lack of bureaucratic structures when compared with other internet standards bodies like ICANN; "bureaucracy is protocol atrophied," Galloway says, and that is where protocol fails, not on its own terms but because it's not permitted to live up to its potential.[56]

Of course, no one with even a passing knowledge of bureaucratic and administrative structure can help but agree with a proposition about its stultifying nature. But I think perhaps this is too easy a dismissal, and I'm interested to open up questions of bureaucracy, what Ben Kafka calls the "powers and failures of paperwork,"[57] to examine its productive tensions, its micro political gestures of affect in our everyday worlds. And one way to do this is to trace the design logics of the RFC series through email development.

Reading the RFC series is not for the fainthearted. It's not simply the sheer volume that can daunt but the intricate level of detail covered within individual posts, making even more impressive Braman's methodological approach in which she found it "essential to read every line of every document."[58] All RFCs contain painstaking explanations of a given document's technical history—outlining reasons for modifications sought, summaries of debates encountered, and procedures to follow for future proposals. In fact, the editorial process has been considered sufficiently rigorous and exhaustive to qualify the series as peer-reviewed scholarly publication for citation purposes.[59] Such administrative practices are built into the memo template of all RFCs, a design choice itself mapped in RFC 5741, which mandates the type and order of headings to appear together with the particular content to be used within each section: that the introduction must contain a paragraph explaining, for example, that the following procedures have been met. In this way we see how managerial cultures and bureaucratic media are central to the development of email. While the content of the RFCs has been comprehensively parsed for clues to the social and technical development of the internet, what goes less remarked is how the formal properties of those RFC texts—the memo format, easy accessibility, and conversational style—have shaped email cultures. I want to suggest, in other words, that the template used by the RFC community laid the foundation for the adoption of the memo header format in email to which we now turn.

Email Header Design

Could it have been different? Why has the memo template remained the industry standard for email since its inception? And just as importantly, what has been the consequence of imagining email as a memo-like media form? To answer these questions, I trace two interrelated critical, material trajectories to consider the development of email headers, memo templates, and interface design located within the flows of bureaucracy. So prosaic and routine to us today, the email header format is in fact the result of intense technical, commercial, and institutional disagreement about its defining protocols. When articulated through a set of RFC documents during the 1970s, efforts were made to establish compatibility between different computer systems and email programs. At stake was how to agree on the format, syntax, and volume of information contained in the email headers, and at times the process of the RFC circulation itself seemed to be getting in the way of arriving at these decisions. Discussions between the various industry and government groups would break down because those responsible for eliciting feedback on the specifications were unfamiliar with the procedures for publishing and circulating the RFC paperwork. To many, the administrative communication process had seemed "chaotic," which meant there were doubts that adequate input from the design community had been sought on how the scope of these protocols should develop.[60] To others it was the reverse. Systems of decision-making were *too* formal. In a post to the MsgGroup email list, Jon Postel questioned these processes, suggesting: "Perhaps there is too much emphasis on official and not enough emphasis on best specification so far."[61]

Postel was taking issue with RFC 724, titled "Proposed Official Standard for the Format of ARPA Network Messages," and he wasn't alone. Hotly debated was precisely how much information should be mandated in the header field, with fears expressed that expanding these fields leads to increased levels of email "trash": unwanted information that cost too much to download or print. There was certainly a case to be made for ease of message retrieval, where detailed identifying information provided (such as "cc," timestamps, action to, and message ID) was useful. But in the everyday environment was it really necessary? And in any case, being required to fill in all these fields (only some would be automatically supplied by the email reader) seemed at odds with the informality and immediacy of the medium. As Austin Henderson put it during the discussions:

> Addition of constraints on form (such as picking an "appropriate" subject) can significantly increase the perceived psychic load, and therefore ought not to be introduced without careful thought about the impact such constraints may have upon the "feel," and therefore the use, of the medium.[62]

Let's pause a moment to note the significance of this "psychic load." Right from the beginning of email, then, people worried about information overload and, specifically, the burden to read all messages arriving in their inboxes. Beyond questions of header volume, the engineers and developers were also interested in new functionalities and in the syntax by which these categories would be organized. For a number of years debate centered on the @ sign of the recipient field. During these header protocol discussions it was suggested that instead of the @ sign appearing only once as is recognized today, the syntax would be designated thus: FriendlyUser@hosta@local-net1@major-netq.[63] How bizarre this appears to contemporary eyes, how it looks nothing like an email address, is testament to the processes by which we become habituated to technology. Commercial and proprietary technical factors were also in play. Standards that the US-government-funded research consultancy Bolt Beranek and Newman (BBN) issued were often unreadable by programs in use elsewhere across the network, and the negotiations were hard fought about which specifications might prevail.[64]

What wasn't in contention, however, was adopting the memo template itself. Such a format had been used informally through the fundamental specifications of three fields—From, Subject, and Date—with each punctuated by colons and the end of the header section, and containing a single blank line.[65] In RFC 724 this template was implemented with the suggestion that email should use "a general 'memo' framework," a form that "severely constrains document tone and appearance and is primarily useful for most intra-organization communications and relatively structured inter-organization communication." While determining that ASCII text would be standard, the way was left open for delivering a "more robust environment" in the future that would "allow for multi-font, multi-color, multi-dimension encoding of information."[66] Underscoring the symbolic and material significance of the memo as constitutive of bureaucratic life and labor are the references within the RFC header series to secretaries and committees. As is conventional with standards literature, illustrative examples of messages are provided to demonstrate certain affordances of the applications proposed. Various scenarios are modeled in which a secretary might log in to send messages on behalf of their manager and to show the syntax of replies:

> George Jones asks his secretary (Secy at Host) to send a message for him in his capacity as Group. He wants his secretary to handle all replies.
>
> From: George Jones <Group at Host>
> Sender: Secy at Host
> Reply-to: Secy at Host[67]

Such models continue to be drawn from in the RFC header protocols up to the most recent RFC 5322 "Internet Message Format" of 2008.[68]

I'd like to make a couple of points about the memo format of email to close this section. First, as JoAnne Yates has so meticulously shown, the memo is a particular form of organizational media that fundamentally shapes labor practices and inaugurates new forms of managerial style understood as "control through communication."[69] Emerging in the early twentieth century, the memo marks a shift in business structures from methods that are "ad hoc," and operating within small factories owned by individuals, to large-scale, board-run, dispersed companies requiring a "systematic management" approach. This developing style of organizational management is defined by its "documentary impulse" and is shaped by the design of new office technologies including the typewriter and vertical filing cabinet. The memo reconfigures older forms of business communication and responds to a growing emphasis on corporate efficiency and hierarchy by introducing brevity in composition, ease of document circulation, and innovation in archival method. Where the business letter was directed externally, the memo is one of the first internal genres of organizational communication. Because memos were destined only for internal circulation, they could dispense with the formalities of traditional business letters. Rather than the complicated requirements of correct address and salutation, a memo standardized these elements for ease of composition and rendered communication routine and predictable.[70]

Second, the design choice of the memo format for email affords particular kinds of interaction: casual in tone yet programmed. When thinking through the increasingly automated processes of digital life, I suggest the connections between early twentieth-century office technologies, memo formats, and email is useful to acknowledge.

Conclusions

Part of the challenge for understanding email is, paradoxically, its familiarity. So routine are its processes, so banal is its operation, it is sometimes difficult to get a proper grasp or perspective on its scope. In order to begin the necessary mapping work, I call the first section of the book "Histories and Landscapes." Its remit is to provide some temporal and spatial coordinates from which to begin telling the story (or stories) of email. In this chapter about the origins of email, I have attempted to give an overview of the key moments of email history and, importantly, outline how these developments have contributed to the look and feel of contemporary email. The early days of email initially involved the exchange of electronic message just within a single computer and then later between different machines. At this time the @ sign made its way into computer history when Tomlinson chose it to define the syntax of email addresses.

While email was gaining traction within the research and scientific communities it caught the attention of the commercial sector as the United States Postal Service saw that new potential revenue streams might emerge. During the late 1970s, news of these USPS proposals traveled quickly through the computer and engineering networks. Some, like Ted Nelson, worried about the future of email research and development if it was taken over by Washington. As he put it during the initial planning stages, "What the Postal Service wants to do might narrow and restrict what you can do in the future with your computer."[71] In one sense, traffic between email and the postal service has persisted for decades through the graphic iconography still deployed in most email client software—its visual language of "mailboxes," "stamps," and "envelopes." But in the early 1980s the economic connection was not to hold because the short-lived E-COM service never gained a foothold in the email provider market. The postal service would still play a role, however, in the expansion of email. Those RFCs that documented many of the design processes for email protocols were actually first circulated through the post before going online in subsequent years.[72]

Reading the RFCs reveals that the technical decisions taken, what fields to include in the email header, for example, were often inflected by social and political concerns. Arguments about email standards took place within a sector made of different industry, academic, and government stakeholders. In addition to the challenges that competing interests might bring to the technical decision-making process was how the RFC itself operated as an administrative and distribution system. Discussions about how to modify email actually occur via email. As I will show throughout this book, one of the intriguing elements of email (as media) is how it often acts as both (or either) content and form. Recall in the introduction that we saw how email functions as a research method: scholarly papers are frequently "about" email only to the extent by which the data collection has been carried out via email survey or interview, for example.

One of the enduring legacies of the RFCs concerns their decisions about email header design and, in particular, the choice of the memo template. The RFCs themselves were organized around a memo format, so it is not hard to see the connection linking today's email programs with the administrative forms in use by the scientific community of the time. In tracing a line from the formal and social conventions of the RFCs through the mid-century popularity of the office memo template to contemporary email, I am aware of the slightly speculative nature of this kind of historiographic work. To a certain extent, this chapter has been concerned throughout with the "meta" level of history, questions about what should be included, what claims are considered, and so on. As I argued, the notion of "counter claims" participates in a certain sort of technological history that often privileges "firsts"—"individuals" and "inventions"—over the less sensational, slow, collaborative, and incremental efforts of the many. In the next chapter I focus in much more detail on this metahistorical level by exploring what happens when loud and singular claims are indeed made about the invention of email amid a complex intersection of celebrity, law, and media.

2 "Inventing Email" and Doing Media History

Media history often unfurls along a daunting and unruly path. Every invention comes with the echoes and trails of its precursors; here is the selfie in portraiture, the internet in the telegraph, or the cinema in the early days of rail travel. Unspooling the past, tracing media forms back to their origins, is a task that beguiles and sometimes defeats its narrators. Along the way, critical frameworks and terms offer guidance. "Remediation" is one such framework: a process for refashioning) each new technology by viewing it through its predecessors; or there's media archaeology, which alights on "counter-histories" to illuminate the forgotten moments of emergence.[1] These historical approaches themselves are not without limitations. Often at stake is the attribution of agency. To what extent is it possible to identify individuals as inventors without overlooking the multiple social and economic forces at play? Conversely, how to acknowledge the media materiality of our communication tools, their conditions of production, affordances, and particularities, while still remaining alive to the human embedded labor and social relations?

In the previous chapter I tried to strike a balance between these sometimes competing impulses by looking at the materiality of, for example, the RFC system on the one hand and the people who composed and circulated the RFCs on the other. I attempted to keep at bay the often-blunt instrument of historical excavation that seeks absolute firsts and pioneering inventors. Now however, it is necessary to face some of these questions head-on because it is in this chapter where digital historiography seems particularly enmeshed within social, technical, legal, and economic currents.

Definitions and Staking Claims: V.A. Shiva Ayyadurai

If email definition, development, and expansion seemed a relatively unproblematic story (as explained in the previous chapter), this would change dramatically in 2012 with the emphatic claim from V.A. Shiva Ayyadurai that he, not Ray Tomlinson, invented email. For many readers, this episode should be relegated to a footnote. Some will no doubt feel that to entertain his assertion, to give it any coverage in a formal or historical overview of email, perpetuates a myth that has been roundly and authoritatively condemned by the internet research community. And yet his arrival on the scene, the narrative threads that knit together enthralling conspiracy theories, "post truth" news production, cultural heritage policy, and celebrity discourse does demand a certain degree of analysis, particularly for scholars of media and communication.

According to Ayyadurai, the real story begins in 1978 when, as a 14-year-old boy, he designed an electronic communication system for the University of Medicine and Dentistry of New Jersey (UMDNJ) where his mother was working as a data analyst. Ayyadurai called the program EMAIL because he had closely modeled it on the university's internal paper-based mail network, and it comprised inboxes, outboxes, folders, memos, and attachments. Adopting the new system, university employees then began to exchange messages using many of the features recognizable today as email including *to*, *from*, and *forward* together with address book and attachment provisions. In the foreword to Ayyadurai's 2013 book *The Email Revolution: Unleashing the Power to Connect*, the then director of the Computer Science Laboratory, Les Michelson, explained that he had hired Ayyadurai as a Research Fellow despite his youth. "Shiva was given a challenge," he wrote, to "create a computer program that would be the electronic version of UMDNJ's interoffice, interorganizational paper-based mail system."[2] Communication at the university was conducted across a number of organizational units and campuses through the central tool of the memo. As we saw in the previous chapter, the memo was a dominant form of organizational interaction at the time, providing a perfect model for email architecture to emulate because of its one-to-one and one-to-many format.[3]

Ayyadurai's first task was to embark on a scoping study of the university's communication infrastructure: its volume of memos, their purpose, paths of circulation, and means of transport. Key to automating the current memo system was ease of use. Ayyadurai was particularly insistent about the usability of his invention; because computer scientists were not its target audience, it had to be intuitive.[4] After successfully road testing the new program across the university computer system, serving some 500 users,

Ayyadurai applied for copyright protection in 1981 by submitting the text of the program and the user's manual. His application for a program called "EMAIL: Computer Program for Electronic Mail System" was registered in 1982 by the US Copyright Office.[5] In 1980, a local newspaper published a story about his achievement; this is the earliest independent reference to his email system.[6] Ayyadurai has regularly shifted the "invention" date in his various press releases, books, and websites as occurring somewhere between 1978 and 1980. Throughout this chapter I will use his 1978 date, as he claims, with the proviso that the earliest verifiable corroboration remains as 1980, the date of the news item.

During the intervening years email became a global phenomenon. Ayyadurai meanwhile gained a number of postgraduate qualifications from MIT including a doctorate in biological engineering. While at MIT he studied under the well-known linguist Noam Chomsky, a fact that would become significant as the events played out in early 2012. Before this time Ayyadurai worked closely with the USPS as a consultant and active postal system reformer. According to Ayyadurai's account in *The Email Revolution*, he had been advising USPS senior officials as far back as 1997 that they should offer an email service but, as he argued, this had "fallen on deaf ears."[7] In fact, at times, he worried that his own software development with email had contributed to its demise.[8] Interviewed by *Time* in 2011 he was asked, "Do you think email is killing the Postal Service?" Ayyadurai responded with the observation that time and again the USPS had missed business opportunities because of a fundamental misunderstanding about the function of email. As he explained, it's the "electrification of letters—it's not just messaging," meaning the USPS should take over areas like bill paying that previously were the province of letters.[9] As we will see, he has not been alone in such recommendations, as the USPS itself has often expressed similar views.

Ayyadurai has collaborated with the USPS on a number of projects, including a conference run by MIT in 2012 that addressed the "uncertain future" of the post office. The event, which Ayyadurai chaired, drew together USPS business strategists with policy makers. One of the key discussion points was security and control of information. How could the USPS contribute to urgent discussions of "big data," its production, breaches, and access? USPS Inspector General David Williams and Ayyadurai both suggested that innovation opportunities and new revenue streams for the ailing organization lay with providing services and applications that delivered greater security through email. Adopting secure email had worked well with the other national post offices of Israel, New Zealand, and Germany.[10]

Ayyadurai began to attract attention within internet and technology research circles during this period partly because the Smithsonian National Museum of American History recognized his contribution to the field by archiving his software documentation, tapes with the original Fortran code, the copyright certificate, and the user manual. It also acquired some of the presentations that Ayyadurai had given to the staff at UMDNJ as the program was introduced.[11] Reporting on the Smithsonian's decision, the *Washington Post* ran a story in February 2012 under the headline "V.A. Shiva Ayyadurai: Inventor of E-mail Honored by Smithsonian."[12] And here's where history pushed back.

Publications and News Stories: Responses to Claims of Invention

Taking to their blogs, listservs, and the comment fields of popular technology publications, members of the internet research community expressed outrage at the article. One significant problem has been the moniker "inventor of email," which journalists (for the most past) applied to Ayyadurai because of the many "vanity" domain names he has registered, including *inventorofemail.com, historyofemail.com,* and *emailinventor.com.* It was this branding strategy, rather than the Smithsonian decision, per se, to archive Ayyadurai's work, that prompted the vociferous backlash.

In particular, the computer historian Thomas Haigh has been persistent and methodical in his repudiation of Ayyadurai's claims. Haigh is currently an associate professor of history at the University of Wisconsin and has served as chair of the Special Interest Group for Computers, Information and Society (SIGCIS), a leading international community of historians, information technology scholars, social scientists, and industry practitioners operating in conjunction with its parent organization, the Society for the History of Technology. Established in the mid-1980s SIGCIS past chairs have included the high-profile internet historians Janet Abbate and Paul Ceruzzi, and its mailing list has over 500 subscribers. Given its membership of scholars, technology experts, and cultural heritage practitioners it is not surprising that news of the freshly minted "inventor of email" spread quickly through its channels. In one of the first posts to dispute reports about the beginning of email, Haigh pointed out the illogical nature of inventing something that is clearly already in existence.

For Haigh, the facts of email history are indisputable and predate Ayyadurai's work. The initial stage, discussed in chapter 1, occurred in the late 1960s with mainframe mail. As Haigh explains, "MIT is a strong contender for the first place where this happened." Ray Tomlinson further developed these functionalities in 1971 as networked mail was exchanged between computers. So, what is it exactly that "Ayyadurai is supposed to have done?" wonders Haigh. All that can be proved is that Ayyadurai owns the copyright for a specific software program rather than for the actual invention. Moreover, this registration does not establish novelty because that is the province of patent law. As Haigh argues, the original *Washington Post* piece was

> confusing copyright protection with patent protection, and implying that he would only have copyright on a program he created if it was the first of its kind. I could write a program called "OPERATING SYSTEM" tomorrow and hold the copyright, but it wouldn't mean I invented operating systems.[13]

Haigh's view was supported by many others who had worked on the original systems at MIT and BBN including Dave Crocker, a key consultant with the Internet Engineering Task Force. In a letter to the Smithsonian, he outlined some of the central difficulties apparent in Ayyadurai's claim. For example, Crocker made available documentation to show the prior existence of the *to*, *from*, and *subject* fields, functionalities which Ayyadurai had asserted were invented with his program. Since 1971, wrote Crocker, the "service has actually been in continuous operation."[14] Crocker also pointed out that he and Tomlinson were awarded the IEEE 2004 prize for their work in the early 1970s on developing email software and standards.[15] In response to the widespread and energetic discussion, Emi Kolawole, the author of the initial *Washington Post* article, issued the following correction under a revised headline that read "Smithsonian Acquires Documents from Inventor of 'EMAIL' Program":

> A previous version of this article incorrectly referred to V.A. Shiva Ayyadurai as the inventor of electronic messaging. This version has been corrected. The previous, online version of this story also incorrectly cited Ayyadurai's invention as containing, "The lines of code that produced the first 'bcc,' 'cc,' 'to' and 'from' fields." These features were outlined in earlier documentation separate from Ayyadurai's work. The original headline also erroneously implied that Ayyadurai had been "honored by [the] Smithsonian" as the "inventor of e-mail." Dr. Ayyadurai was not honored for inventing electronic messaging. The Smithsonian National Museum of American History incorporated the paperwork documenting the creation of his program into their collection.[16]

For many, however, this correction did not go far enough and worse, it still seemed to accept the broad premise of Ayyadurai's claims to invention. Populating the comments field of the *Washington Post* article were the views of software developers and engineers, some who had actually authored a number of the key RFCs about headers and had been involved in the early development of email. These commenters took issue with the self-branding strategies of Ayyadurai that functioned to eclipse any other contributions made to the steady growth of email from the 1960s. Perhaps in recognition of the overwhelmingly negative tone of the reader feedback—nearly all of the 108 comments criticize Kolawole's article, despite its subsequent correction, and then call for stronger clarification—and together with letters of complaint sent directly to him, the *Washington Post* ombudsman, Patrick Pexton, felt it necessary to intervene. His response, however, seemed only to inflame the situation. In unmistakably dismissive tones, Pexton opened his column by posing the rhetorical question, "Who invented e-mail? Crikey, I don't know. Maybe Al Gore."[17] Clearly defending the original piece, Pexton chastised the readers for misunderstanding the fundamental processes of newsgathering, again asking rhetorically: "Why is it that scientists, academics, and some readers, think that journalists and newspapers should be like academic journals and peer review every sentence that appears in print?"[18] A few days later however, Pexton posted a further ombudsman clarification under the headline "Origins of E-mail: My Mea Culpa."[19] Here the *Washington Post* did accept blame for overemphasizing many aspects of the story and in fact admitted to getting other elements factually wrong. Pexton wrote:

> I did a blog post this past Friday that was dismissive, snarky and wrongheaded, and had factual errors too.... I was upset at the harsh nature of some of these e-mails, and they came amid a heavy week of barbs and complaints about Post coverage—some of them merited, some of them not.... But I think it's safe to say that although Ayyadurai is an interesting fellow, and that, as a teenager, he did develop an early electronic messaging system for about 100 users at the University of Medicine and Dentistry of New Jersey and obtained a 1982 copyright for its computer code—he named the program, all uppercase, "EMAIL"—he should not have been called "inventor of e-mail" in the headline. As so many distinguished experts in this field wrote to tell me... Ayyadurai is not the inventor of electronic messaging between computers, what we have all come to call e-mail. Electronic messaging was developed by many hands over many years, and probably began in the early 1960s, possibly as early as 1961, on people using time-shared computers.[20]

As the *Washington Post* notes, on February 23, 2012, the Smithsonian also issued a statement refining its earlier descriptions about the basis on which Ayyadurai's materials were archived. The museum explained that although exchanging "messages through computer systems, what most people call 'email,' predates the work of Ayyadurai," it "found that Ayyadurai's materials served as signposts to several stories about the American experience." Such stories include a complementary focus to the historical studies of the ARPANET's use of email. Instead, "Ayyadurai's story reveals a contrasting approach, focusing on communicating via linked computer terminals in an ordinary office situation," a localized system that linked "only three campuses rather than multiple large institutions." Concluding its statement, the Smithsonian noted Ayyadurai's initiative "was a small enterprise, rather than a big enterprise story."[21]

But the matter did not end there. A few months later a press release in support of Ayyadurai was issued in the name of Noam Chomsky. Key to Chomsky's validation of Ayyadurai's claim was the unethical rewriting of history to erase a voice excluded by what he dubbed "industry insiders." A "deplorable" situation had arisen where the "childish tantrums" of those who believed that by "creating confusion on the case of email," they could "distract attention from the facts." Chomsky added that "there was no controversy here," save for the work of this select group who had a "vested interest" in advancing their particular narrative.[22] An insider group? One could almost call it "a media elite."

But more about that shortly: first it might help to flesh out a little further what Ayyadurai argues is his unique contribution and, further, how he refutes his critics.

Ayyadurai's Specific Innovation and the Conspiracy against Him

Ayyadurai argues that his specific innovation was to provide a "system of interlocking parts" whereas what had been in existence was simply electronic text messages exchanged on the 1960s mainframe computers. For Ayyadurai, this earlier design was "pre email" so he grants to Tomlinson the expansion of these messages from a single machine to those transferred between different hosts. However, these early "primitive" systems did not provide users with functionality like an address book or attachment capacity. Instead, Ayyadurai had created a "full-scale emulation of the interoffice interorganizational paper mail system."[23] He also disputes the authority of the RFC system—the widespread process in use since 1969 by which computing documentation and standards are established—as proof of the invention of email by those in the scientific and defense communities. RFCs are not code, he points out, and what *he* did was write the actual software program, some 50,000 lines in FORTRAN IV code.[24] So it doesn't matter, he argues, whether there are RFCs that discuss the scope, future potential, and technical permutations of electronic messaging, he was the one to do it.

At this juncture it is worth noting that there has been very little dispute concerning Ayyadurai's claim to have created an early email program. Dotted through the relevant articles, comments fields, and lists has been praise for the obvious achievement of someone so young. And for many it is the instantly recognizable ordinariness of a university or lab setting, the memos, the organizational structure, the banality of meetings and workflows, that at times makes the story ring true. But the problem for many is the extrapolation that Ayyadurai has then made. As Larry Tesler points out, posting to Dave Farber's "Interesting People" list when the various news items first began to be shared:

> I do not wish to detract from Shiva's impressive accomplishments at age 14, which included user studies, architecture and engineering for an early email system commissioned by the University of Medicine and Dentistry of New Jersey.... But, like your other readers who commented on recent press reports, I think it's important to view his 1978–79 project in the context of other work during the 1970s that directly influenced the email we know today.[25]

It's clear, however, that Ayyadurai seems never to have been interested in embracing a view of technological innovation that sees measured, incremental steps taken at various points or a sense of collaborative contributions unfolding over time. Instead he argues that a systematic, institutionalized campaign by the "industry insider clique" has been waged to discredit him, a group comprised of questionable historians (that term is often enclosed within scare quotes) and the IT company Raytheon BBN Technologies. Writing in the third person, he explains that

> industry insiders, supported by SIGCIS "historians" Ray Tomlinson, BBN supporters, and ex-BBN employees continued to perpetuate a false history of email by discrediting Ayyadurai's invention. They used revisionism and confusion to redefine and misuse the term email. Through these efforts they re-declared Tomlinson, and thereby the BBN brand, as the "inventor of email"... in popular press releases between April 24–26, 2012.[26]

Ayyadurai is impassioned—some would say hyperbolic—about his assertions. Nonetheless, the fact that BBN has declared its invention of email as part of its own branding exercises has at times provoked the ire of those from the very same early internet communities. Writing in response to the news of Ayyadurai's claims, Dave Walden points out that during the period he worked at the computing company many had "tried to stop BBN people who didn't know any better from saying that BBN or Ray Tomlinson invented email." What provoked their critique of BBN was the desire to be "fair to the people who prior to Ray's demonstration of networked email already had what today we call email running within single computers," or time-sharing systems.[27] If there was further evidence needed of the complicated weave of history, with its competing and contradictory paths of invention always inflected by both individual desire and socioeconomic forces, this is it. For on some issues, perhaps Ayyadurai has a point: BBN's expansive claims to history could afford a little nuance. Even those who worked there agreed. As a number of leading internet histories have shown, the relation was sometimes testy between BBN and the wider "minority" computer research organizations, with many feeling that BBN wielded too much power in decisions over the development of communications infrastructure.[28] More anecdotally, the nickname "Big Bad Neighbor" was bestowed on the company by Mike Padlipsky and was occasionally invoked on mailing lists to convey some of the tensions that existed between the various stakeholders in the development of ARPANET.[29]

And yet there is something quite overblown—indeed, misplaced—about Ayyadurai's insistence that there exists an obvious, traceable line stretching from Gmail today back to his invention at UMDNJ in the late 1970s. For one thing, as many in the computer history community have asked, where are the journal articles, conference presentations, or technical reports one would expect to see as evidence of impact and peer review in scientific discovery?[30] Without such documentation it is difficult to validate Ayyadurai's claims to the invention of email. This is not to deny that Ayyadurai has some expertise in the email software provider market. One measure of his general impact might be EchoMail, the email management company Ayyadurai launched in 1993 to automate and process large-scale email traffic. His commercial clients included AT&T, Kmart, Nike, and American Express together with servicing the government sector where it coordinated the communication between US senators and their constituents.[31] This aspect of email expertise, at least, is not in dispute with Ayyadurai's commercial success making inroads into the email direct marketing and filtering software industries. Of course, the mere fact that he has worked within the broad email provider industry does not of itself justify his claims to invention.

Assessment of Ayyadurai's Claims

The documented evidence demonstrating that email was in widespread use before Ayyadurai's claim of 1978 is comprehensive and irrefutable. As mentioned, the RFC series represents a time-stamped, publicly accessible record of the development of email systems from the 1960s to 1980s (and of course RFCs are still in use as improving protocols for email). A few specific and significant examples of those RFCs that outline early email technology—RFC 561 (titled "Standardizing Network Mail Headers" and published in 1973), RFC 680 (titled "Message Transmission Protocol" and published in 1975), and RFC 733 (titled "Standard for the Format of ARPA Network Text Messages" and published in 1977)—all clearly document that the fields of *to*, *from*, and *cc* were being comprehensively defined. In RFC 524 (titled "A Proposed Mail Protocol" and published in 1973) there is early consideration of delivery receipts; this RFC is also noteworthy for beginning to establish mail as a stand-alone protocol separate from File Transfer Protocol (FTP). Finally, RFC 706 (titled "On the Junk Mail Problem" and published in 1975) is, as the name suggests, recognition of the existence of spam or unwanted email content on the network. As has been widely discussed, the first documented case of commercial spam dates back to 1978 when an employee of the Digital Equipment Corporation (DEC) flooded the network with advertising material sent via email.[32]

It should be noted that while RFCs are vital for establishing timelines for the introduction or development of certain applications as I have done above, they are also an invaluable record of popular yet informal practices and norms of the day. In other words, RFCs capture what was already going on. Their purpose is often to formalize and standardize technical and social solutions, sometimes referred to as "hacks," that were in widespread use but in need of standardization. This everyday use of RFCs in the development of email seems always to elude Ayyadurai. In addition to the RFCs as public-record evidence of email prior to 1978, one of the first trade publications devoted to email was launched in 1977 called *Electronic Mail and Message*

Systems. The Computer History Museum, which holds archived copies of the document, describes this as a "twice monthly newsletter covering technology, user, product, and legislative trends in graphic, record, and microcomputer communications."[33]

Doubtless Ayyadurai would respond that such "evidence" shows only the creation of rudimentary electronic *text messaging*. As I explained above, what Ayyadurai claims to have invented is a system of "interlocking parts," and he places great emphasis on its similarity with the paper-based postal system. In Ayyadurai's words from his website: "Email is the electronic version of the interoffice, inter-organizational paper-based mail system. Email is not simply the exchange of text messages."[34]

There are two quick points to make here. The first is, as Thomas Haigh has explained through his meticulous version of controlled documentation of Ayyadurai's claims over the last seven years, Ayyadurai has himself regularly changed the definition of what counts as email in order to retrospectively claim invention. For example, he dismisses early email programs like Xerox PARC's Laurel in use in the late 1970s because it relies on a server rather than acting as a stand-alone client. Yet he then includes contemporary email programs like Eudora and Outlook in his timelines as proof of his influence, although these are also based on a client-server model. That is, any applications featured on early email systems must be defined as "text messages," and in post-1980 typologies will be included as email.[35]

The second point to note concerns Ayyadurai's insistence that only he and he alone has made an explicit connection between email and the paper-based postal service. As part of the argument, his website contains quotes from early ARPA and BBN computer scientists apparently dismissing the idea that email was designed to mimic the postal service. But as I outlined in chapter 1, the postal service itself, both in the US and internationally, had already noticed the similarities and potential business opportunities offered by email. In 1980, the USPS held a conference in Washington called "The Electronic Mail and Message Conference," which floated ideas that would eventually become Electronic Computer-Originated Mail (ECOM or E-Com). Two years earlier, in 1978, the USPS had presented its plans to the Postal Rate Commission for approval. Not only is this clear proof that these material and conceptual links between the postal service and email were already well in train by 1978, but it also shows that the computer science communities themselves were making these connections. As I also noted in chapter 1, news of the USPS's plans made the rounds on mailing lists at that time. In 1977, for example, the MsgGroup list, one of the earliest mailing lists, began discussing the possible encroachment of the USPS into the area of electronic mail. Einar Stefferud wrote prophetically about it:

> It is my opinion that the advent of the "home" computer, which will also become the "personal" computer for both home and office, will make the perfect "mailbox" for electronic mail, and that the part of the USPS which does home delivery is one of the key parts to be bypassed by electronic mail.[36]

Stefferud was here responding to a news article circulated through the list; as explained by the original poster, it was about "how the post office is worried that they might get ripped off if everyone goes the ARPAnet mail route in the future." Jeffrey Mills, who wrote an Associated Press article dated January 4, 1976, opens with the following statement: "The Postal Service is beginning to worry about competition from electronic communication systems which threaten to make mailmen obsolete."[37]

So, if the post office connection cannot hold for establishing Ayyadurai's unique contribution, then what can be said about his role in email history? As I have explained throughout the chapter, there has been little disagreement in computer history communities that during the period from the late 1970s to early 1980s Ayyadurai wrote a creative and useful email program. But as to his invention of the entire system of email, that has by now been thoroughly discredited.

Ayyadurai's Legal Actions: *Gawker*

At this point let me note again why I've chosen to include this contour of the internet landscape in my book. For those who believe Ayyadurai's only real expertise lies in "spin," and that the real driver in this version of history is Larry Weber, the PR manager he has commissioned, the work I undertake here to include it simply replicates a myth. But as historians have taught us, what happens in the margins often deserves attention.[38] I consider this point below when I explore the lessons to be drawn from this event for media history and to what extent it should be considered using terms like "marginalized" or "neglected."

But now we need to return to the unfolding story as it erupted once more with Ray Tomlinson's death on March 5, 2016. The announcement prompted discussions across mainstream and social media about Tomlinson's historical and technological achievements. It was hardly a shock, then, when Ayyadurai rejoined the conversation. In a series of outspoken tweets he once again staked his claim that it was he, not Tomlinson, who invented email. And for him the oversight was borne out of race and corruption. Not mincing his words he stated: "I'm the low-caste, dark-skinned, Indian, who DID invent #email. Not Raytheon, who profits for war & death. Their mascot Tomlinson dies a liar."[39] Ayyadurai also felt that the snub represented a particular problem with the US mainstream media: "Indian media sharing truth.... Total BLACKout in US media protecting fraud Raytheon/Tomlinson lies that EMAIL='@'"[40] He followed this up with a number of tweets aimed at CNN, Mashable, and the *Washington Post* in which he demanded "Are you going to continue racist propaganda that Tomlinson invented email. How much AD $'s do you get from Raytheon?!"[41]

Ayyadurai's condemnation of these US media outlets soon stretched beyond Twitter. A few months later he launched a lawsuit aimed at the gossip blog site *Gawker* and one of its subsidiary publications, *Gizmodo*. Similar legal action was taken by the wrestler Hulk Hogan and a number of other litigants, with all claimants represented by the prominent entertainment lawyer Charles Harder. The basis of Ayyadurai's litigation was defamation; he alleged that two stories by senior staff writer Sam Biddle had labeled him a fraud and a fake, the first in 2012 under the headline "Corruption, Lies, and Death Threats: The Crazy Story of the Man Who Pretended to Invent E-mail."[42] These legal actions gained poignant media currency because the prominent venture capitalist Peter Thiel was thought to have funded them. A cofounder of PayPal, a Facebook board member, and a vocal Trump supporter, this libertarian billionaire had himself harbored a long-running grievance against Gawker Media. In 2007 the blog posted an exclusive report with the headline "Peter Thiel Is Totally Gay, People" and claimed in the article "what no one ever says out loud: Thiel is gay."[43] While Thiel took no immediate action, many saw him biding his time until he could seek revenge for *Gawker*'s act, which he viewed as a "unique and incredibly damaging way of getting attention by bullying people" when clearly there was "no connection with the public interest." His retaliation—although he cast it as being "less about revenge and more about specific deterrence"—took the form of financial backing for the invasion of privacy case brought by Terry Bollea (professionally known as Hulk Hogan) against *Gawker* for sex tapes of the wrestler it had published in 2012.[44] Initially the terms of this funding had been kept secret from the public and, in particular, the jury hearing the case, until the *New York Times* revealed that Thiel had paid approximately $10 million in order for the litigation to proceed.

And proceed it did. Following a two-week trial in March 2016, Bollea was awarded $140 million in damages, with another $10 million to be paid personally by *Gawker*'s founder, Nick Denton. *Gawker* appealed the decision, but in May 2016 a Florida judge dismissed both their request for a new trial and a reduction of the jury's award of damages.[45] The case was later settled, and the subsequent appeal dropped by *Gawker*, for $31 million. Ayyadurai, meanwhile, reached a $750,000 settlement in his defamation action for the two *Gizmodo* pieces, the second of which (published in 2014) included snide references to his wife, Fran Drescher, of the long-running 1990s TV comedy program *The Nanny*. It read:

> Fran, you f***ed up! You had two years to discover that this guy is basically a big fake. And now look at you: stuck with the guy, doomed to nod along to his cyber-lies for a lifetime, as you grow old together and reminisce about things you didn't invent.[46]

Ayyadurai had sought compensation for the personal and professional injury he endured as his business reputation was sullied through these publications. His contract lecturing position at MIT had been rescinded, he argued, because of the negative press generated, and he had been forbidden to talk about email by MIT professors at the USPS event discussed earlier, which he had organized. He also claimed that funding had been withdrawn for his MIT-associated research center called the Email Lab.[47]

Gawker's Tabloid History

As a result of these legal decisions, Gawker Media filed for bankruptcy in June 2016. The company was subsequently bought by Univision Communications for $135 million and rebranded as the Gizmodo Media Group. In a statement about the sale and its legal basis, Denton lamented: "After four years of litigation funded by a billionaire with a grudge going back even further, a settlement has been reached. The saga is over." For those of us with even a passing knowledge of *Gawker*'s tabloid history it is a little difficult to feel much sympathy. Recall, for example, the outing of Condé Nast's chief financial officer in 2015. That happened when *Gawker* published a story about rival publisher David Geithner, who had apparently arranged for the services of a male escort. When Geithner decided not to pursue the transaction, the escort went to *Gawker*; the site happily published all the details including intensely candid text messages exchanged between the two men.[48] Finally, amid accusations of "gay shaming," blackmail, and extortion—actions all directed at a person not in the public eye, as many noted—*Gawker* withdrew the piece and Denton apologized.[49]

This topped off a sustained critique that *Gawker* had waged for years against its competitor. In posts tagged "Conde Nastiness," writers would joke, disclose, and gossip about its rival colleagues in the media world. In 2010, there was the misogynist attack that revealed lurid details concerning a supposed sexual encounter between the article's author and the former Republican candidate for senator of Delaware, Christine O'Donnell.[50] So abusive was the piece that the National Organization for Women issued a statement condemning *Gawker* for "public sexual harassment," and adding that, "like all sexual harassment, it targets not only O'Donnell, but all women contemplating stepping into the public sphere."[51] Further back there was the Gawker Stalker, the web application launched in 2006 that utilized Google maps to pinpoint celebrity sightings in real time. Its practices were often criticized for promoting and normalizing everyday surveillance.

Removing Articles to Rewrite History

Following the legal action, Univision, the new owners of Gawker Media, removed a number of articles targeted in the lawsuits, including those about Ayyadurai and email history, from all of the publisher's websites. Senior executives of *Gawker* and staff on its associated sites had resisted such a move because it appeared to rewrite history. *Gawker* itself had criticized *BuzzFeed* for bowing to commercial or political pressure from advertisers, investors, and readers to "disappear" certain articles. Writing to its staff, an outgoing senior *Gawker* executive, John Cook, explained that he had remonstrated with Univision and explained why these articles should remain accessible. As he put it:

> Disappearing true posts about public figures simply because they have been targeted by a lawyer who conspired with a vindictive billionaire to destroy this company is an affront to the very editorial ethos that has made us successful enough to be worth acquiring.[52]

All the posts, however, were removed. In their place a line appeared bearing the explanation, "This story is no longer available as it is the subject of pending litigation against the prior owners of this site.'[53] Internet historians also condemned the move. Katie Hafner, coauthor of the widely cited generalist history of the internet titled *Where Wizards Stay Up Late*, commented it was an "appalling" legal decision in which Ayyadurai had "managed to make money off claims that appear to be misleading."[54] Key to the disquiet was the way in which *Gawker* was "removing history."[55] These news articles had provided crucial facts to refute Ayyadurai's claims, and with this takedown the evidence vanished leaving in its wake only bogus assertions, now in a curious sense perhaps validated by an apparent "legal win." Future historians would get a skewed picture of the real story of email.

Ayyadurai's Legal Actions: *Techdirt*

Where tabloid intrusion could form a narrative context to understand these various lawsuits launched against *Gawker*, the next legal action brought by Ayyadurai was somewhat different. In January 2017, Ayyadurai sued the technology blog *Techdirt* and its founder Mike Masnik for $15 million in defamation damages. At stake were a series of 14 articles published by *Techdirt* between September 2014 and November 2016 that disputed Ayyadurai's claims to have invented email.[56] *Techdirt* wrote many of these posts in response to the initial news stories (published by the *Washington Post* and the *Huffington Post*) featuring Ayyadurai's claims. One such headline identified as defamatory by the lawsuit read "Why Is Huffington Post Running a Multi-Part Series to Promote the Lies of a Guy Who Pretended to Invent Email?" *Techdirt*'s use of the phrases "fake email inventor," "bogus claim," and "guy who pretends he invented email" provided further examples of potential defamation argued by Ayyadurai's legal team in their complaint.[57]

To a certain extent, this case shares aspects with the one Ayyadurai had brought against *Gawker* and *Gizmodo* because both had argued that media articles had been libelous. But it's important to note that Ayyadurai did not "win" his case against *Gawker* in so far as the facts were considered by a court. Unlike Bollea, whose successful litigation was due in large part to the jury's "profound disenchantment" and disgust at *Gawker's* tabloid journalism,[58] Ayyadurai received his settlement because *Gawker*, at that stage in bankruptcy, needed to resolve outstanding liabilities in order to finalize the sale of the company to Univision Communications.[59] In other words, Ayyadurai's claims about his invention of email had not so far been assessed legally. Now, however, with the *Techdirt* case the very definition of terms like "email" and "inventor" were being considered. Unfortunately for Ayyadurai, on September 6, 2017, his defamation case against *Techdirt* was dismissed. In his decision, the judge echoed what many in the computer history community had been saying for some time. Ayyadurai may have created an email program but could not lay claim to the invention of email. Subsequently, the criticisms from *Techdirt* could not be classed as defamatory and were protected speech. As Justice Saylor found:

> The articles at issue do not dispute that plaintiff created an e-mail system. Rather, they dispute whether plaintiff should properly be characterized as the inventor of e-mail based on that creation. Accordingly, it is not clear that the allegations in the complaint are sufficient to show that the statements at issue are false. In any event, even assuming that the allegations of falsity are sufficient, the challenged statements are nonetheless protected under the First Amendment.[60]

Following the court case, Ayyadurai announced his intention to appeal the decision.[61] *Techdirt* has also indicated that it will cross appeal one aspect of the ruling that had denied hearing the case under California's so-called anti-SLAPP law.[62] This legislation prevents legal action—one considered to be a "strategic lawsuit against public participation" or SLAPP—that is taken to silence and intimidate critics by exhausting a defendant's financial or emotional resources until they withdraw their criticism.[63] If *Techdirt* is successful in this arm of their claim, Ayyadurai will have to pay legal costs. Meanwhile, after two years of speculation that Peter Thiel would buy the remaining assets of Gawker Media, in September 2018 it was announced that the Bustle Digital Group had bought the company for $1.35 million, with plans to relaunch the celebrity gossip site in 2019.[64]

Conclusions

What can the tangle of media law, celebrity, journalism practice, and the public domain tell us about email history? Those looking for historical accuracy would say "very little." Ayyadurai's claims have been thoroughly discredited by the broad computer science, communications, and internet sectors. This dismissal nevertheless leaves open for discussion his wider significance within a book about email and the everyday. As mentioned, I have deliberated about the wisdom of including Ayyadurai in the history of email. Clearly, there are major indisputable facts on the historical record which problematize his claims of invention. But at the same time, isn't this event now part of history? Erasing the past works both ways. So, I was curious to learn what internet historians have decided in similar situations.

For leading internet scholar Janet Abbate, the historical record is unambiguous. Ayyadurai's claims to the invention of email have been comprehensively debunked. Although in a study devoted solely to the history of email, the inclusion of his assertions would be a "distraction," Abbate could see value in an approach that took a metahistorical view, widening the scope to explore what it "means to say that someone invented email, and about the politics of historical claims," and in so doing "could raise some interesting historiographical issues."[65]

In this chapter I have explored some of these metahistorical issues by looking at a range of cultural, legal, and political currents that have shaped a moment in the history of email. When trying to understand this moment I was often tempted to apply a frame that looks to the margins of traditional or mainstream accounts of history. Such analyses are well established across a number of distinct yet historically attuned fields often utilized by media and communication research; these include gender studies, science and technology studies (STS), the study of nonhuman agency, media archaeology, and postcolonialism.[66] For these approaches the marginal excavation work will often involve shedding light on a hitherto invisible historical figure or artifact to show how institutional, economic, and social forces have functioned to obscure its relative significance. Underpinning many of these analyses is the recognition that history is embedded in its systems of representation and modes of storytelling. Rather than simply lying dormant awaiting the historian's gaze, the narrative organization of events, the writing of history, is discursive; it actively produces meaning and constructs the reception of a given event into the future. For scholars of media history, the material and symbolic domains present a singular example of this wider point about modes of historical representation. As Lisa Gitelman puts it:

> If history is a term that means both what happened in the past and the varied practices of representing that past, then media are historical at several different levels.... A photograph, for instance, offers a two-dimensional, visual representation of its subject, but it also stands uniquely as evidence, an index, because that photograph was caused in the moment of the past that it represents.

More generally, much contemporary scholarship on how to investigate the past has been inspired in large part by the historiographical contributions of Hayden White. Viewing history as constitutive of its form means being alert to ellipses and silences—who is left out or what remains unsaid about a particular historical moment—together with the use of narrative and other rhetorical devices such as the figures of "hero" and "villain" or the generic categories of satire, romance, tragedy, and comedy.[67] It is important to note that the reach of such analyses is not limited to the confines of academia. Popular accounts of scientific development and technological history regularly draw upon similar interpretive tropes. Alone in the lab, the genius inventor labors away by himself (the figure is most frequently male) perhaps shunning the spotlight or conversely having it stolen from him until the historical writer, historian or journalist, "discovers" his endeavors.[68]

This is certainly how Ayyadurai has presented himself; an outsider who is overlooked and excluded from the annals of history, a stance regularly voiced on his Twitter account: "Stop the racist lies. Raytheon and their mascot Tomlinson didn't invent email, except invent lies & racist propaganda to hide facts."[69] But this representation is difficult to square with someone who has sufficient power to mobilize his celebrity and political connections, first in a PR campaign, followed by litigation, and later includes running for the US Senate. In February 2017, Ayyadurai announced his intention to stand as a Republican against Democratic incumbent Elizabeth Warren in the state of Massachusetts but later decided to run as an independent.[70] As part of his platform Ayyadurai coined the slightly perplexing slogan "only a real Indian can beat a fake Indian" (a reference to Warren's controversial claim of Native American heritage).[71] Ayyadurai gained only 3.4 percent of the vote in the midterm elections held in November 2018 but has advertised on his website his intention to run in the 2020 elections. Because he has the resources to campaign and litigate it is very hard to see Ayyadurai as an "underdog" despite his energetic promulgation of such a persona. What seems never to be in contention for Ayyadurai is his self-proclaimed inventor status. As I have discussed, while recognizing that the narrative construction of history offers productive lines of inquiry and analysis, it can also become reductive, closing down rather than opening up the past. Sole inventors or lone pioneers, together with the language of "firsts," "only," "never again," or "never before," participate in a stark, dichotomous view of history. Either it's this or that. Either email was invented by this person or that person. The lived reality of technological development is much messier and nuanced than stories told with "firsts." And in Ayyadurai's case the inventor trope has had the unfortunate consequence of seeming to deny many of the contributions made by his "fellow" computer scientists, engineers, programmers, and students over the last few decades.

My inclusion of Ayyadurai's claims in a book on email and the everyday, therefore, serves three interrelated purposes. First, it operates as a case study about the inadequacy and shortcomings of historical analyses that rely on the stark language of "invention" rather than on the everyday moments of evolution and collaboration. Second, Ayyadurai is a compelling example of the dance between the banal and the extraordinary that must be considered in studies of the everyday. In the opening pages of this book I explained that one of the curious conditions of studying the everyday is that doing so also reveals the extraordinary. The more focus one exerts on a cultural phenomenon, the odder it seems (a corollary of the well-known linguistic experience of saying the same word repeatedly until it loses all meaning). While his claims have been disproved, the singularity of the event, its drama, is nevertheless part of a history of email. And finally, Ayyadurai's "event" also shows how the everyday is itself a powerful narrative device for telling history. At the core of Ayyadurai's story, landing at all the instances where it did meet a degree of endorsement, were his descriptions of the banality and everydayness of a suburban university setting in which he had developed his email system. Paradoxically, then, these descriptions of ordinary discovery and innovation managed to confer upon his claims the sense of authenticity that his own hyperbole was unable to achieve.

3 The Email Industry

Before the terms "reach" and "engagement" came to dominate how we speak about tracking our online interactions, there were "opens." Just as studying social media analytics reveals patterns of commercial and cultural practice in what we call "engineering sociality,"[1] so too are email metrics illustrative of our everyday socioeconomic relations. In the previous chapters I mapped email history by peering closely at the screen, looking at the protocols by which emails travel to us and then understanding the memo format through which they are read. Now let's zoom out further and appreciate the wider media ecology created by email. By this I refer to the email service provider industry, a loose knit of partially interconnected organizations, technical infrastructures, and software applications. Diverse products and business models constituting a mix of independent, open source initiatives rubbing shoulders with proprietary conglomerates like Gmail inhabit this complex media landscape, but they don't offer the same type of product or service.

Some will design email-marketing campaigns for business while others only deliver free cloud-based email applications. Still others might deal exclusively in the domain of email encryption software. Indeed, the terms used to describe the field are not necessarily employed with precision. "Email service provider" (ESP) could encompass the provision of all the features just outlined, or it may refer more narrowly to the email-marketing sector that is concerned with advertising-campaign strategies involving the maintenance of subscriber lists, data analytics, and content creation. Even then, further distinctions may be drawn between those who provide email distribution (that is, actually sending emails on your behalf), and those who only test and analyze the content. I attempt to bring some specificity to this scoping work, but the blurring boundaries regarding how the terms get put into play is testament to the intersecting and shifting nodes that make up the terrain itself.

This chapter marks the conclusion of part I on histories and landscapes—"landscapes" because to grasp the everyday material impact of email we need to capture its breadth as an economic and social system, its constituent parts, and how users interact with its platforms and services.

To chart this terrain I begin by attempting to capture the size of the market as an industry and look at the different metrics used to calculate its value. In order to get a sense of the "major players" I then explore the email service provider sector in broad brushstrokes, its leading brands and their company structure and impact. It seems to me that the existence of "an email industry" has not been conceived in the same way, for example, as the entertainment business. Where various segments of the media or creative industries are comprehensively investigated by communication studies from political economy, discourse analysis, consumption, or audience perspectives, the same critical frame has not been applied to email. It's true that information technology, software development, or the field of telecommunication more generally receive attention, often through the frame of regulatory or statutory regimes, but viewed as a discrete commercial field, email is underexamined. Although the complex sprawl of the industry is often invisible to us as we battle with yet another notification or sigh at an office "reply all," it actually contributes materially to what we see on our screens and the way we engage with email. After mapping the industry in general terms I then provide a brief case study of the Australian business Fastmail, an email service provider with more than 14 years of experience in the sector. Here I delve into some of the commercial decisions taken by an email provider together with the technical and software design developments that Fastmail has contributed to the area more broadly.

Email marketing represents a significant segment of the email industry. So, I then analyze the algorithms and analytics used in planning, executing, and assessing email-marketing strategies. As we will see, how the sector measures its impact is a process with similar pitfalls and caveats as those experienced when deploying the tools of social media analytics. Although the purpose of email advertising is principally to generate commercial revenue, charities and the not-for-profit sector also leverage email campaigns to great effect. Perhaps one of the most recognizable icons of the email industry is the emoji. Its rendition and interpretation is often complex, with users not always in agreement about its meaning. If email provision is dynamic and mutually constitutive at the commercial and software engineering levels this also occurs at sites of user engagement. In my surveys, I found that very few people limit themselves to a single email program or platform of access. Typically, access and use is dispersed across a complex range of email clients, internet service providers (ISPs), and devices. Gmail is for the IPad, Outlook happens at work on a laptop, and email via phone is for reading but not replying.

The Size of the Email Market

The financial value of the email market as a whole can be calculated using a range of different criteria and processes of segmentation. Some analyses will include only cloud-based email vendors while other reports will limit analysis to enterprise rather than consumer use of email. The methodology used by a particular analytics firm is often subject to proprietary confidentiality, so it is almost impossible to see exactly how certain figures are calculated (in some cases I contacted these companies to request further clarification). But in general terms the email market is currently estimated to be worth about $47 billion annually worldwide.[2] These findings come from a company called the Radicati Group, which includes in its estimations cloud-based email for businesses with products like Google G Suite and Microsoft Office 365; consumer email; and "on premises" or desktop platforms for organizations, including Microsoft Exchange.[3] Radicati reports that the migration to cloud computing from on-premises applications is driving significant change; consequently the market for cloud email in 2020 is estimated to be worth $42.8 billion. They predict that 87 percent of all business email will be cloud based by the end of 2022.[4] Radicati's figures are broadly supported by similar industry data. Statista estimates that by 2024 revenue for cloud-based email for enterprise will increase to $81 billion.[5] Often analyzed as a discrete segment of the broader email industry, the email-marketing sector is currently worth about $4.5 billion worldwide with predictions that it will generate about $22 billion by 2025.[6] Having provided a quick snapshot of market share I now want to outline the major players in the email provision sector.

Email Service Providers

The term "email service provider" is the top-level category to use when mapping the email industry. It encompasses a wide variety of user environments, software applications, hardware architectures, and business models. Indeed, the edit pages of Wikipedia are regularly filled with lively debate about how to capture adequately all the elements of email transport, composition, and delivery. In this section I set out the major players in the field, further elaborating on some of these under subsequent headings. For example, the marketing company Mailchimp is one of the leading names in the email industry and warrants mention in this top-level description. *Forbes* ranks it as no. 11 of their 2018 "Cloud 100" list, an annual index of the top private companies in cloud computing.[7] But it is an email marketing company rather than an email hosting service.

In terms of market share, the leading email service provider is Gmail, a product of Google LLC, itself a subsidiary of parent company Alphabet Inc. According to the web traffic analytics tool Alexa, Google ranks as the most popular website both within the United States and worldwide. Its no. 1 position is calculated by combining average daily visitors and page views.[8] In 2018 Google announced on Twitter that Gmail had reached 1.5 billion users worldwide.[9] This figure represents an increase of half a billion active users over four years.[10] Alphabet employs about 94,000 staff, and in 2017 its annual reported revenue was $110,855 million.[11] As well as its dominance within the email sector, Alphabet is a significant presence in the advertising industry through its applications AdWords and AdSense.

The majority of the company's revenue is generated through advertising, and the enterprise research database MarketLine Advantage states that the "company's performance in advertising business makes it the leading player in the industry."[12] I draw out some of the implications of Alphabet's advertising focus for Gmail below by looking at the way it shapes the field of email analytics. Whereas Google is the market leader for providing the most popular, free (although ad-supported) web-based email product, other key players in the email provision sector are ranked in order of annual revenue and listed with their email clients or platforms: Apple (Apple Mail); Verizon Communications Inc. (AOL, Yahoo!); Samsung Electronics (Samsung Galaxy Email); Amazon (Amazon Simple Email Service); and Microsoft (Outlook, Office 365, Hotmail).[13] Alongside these major household brands are some lesser known companies who are nevertheless key providers offering email hosting services for both consumers and the small-to-medium enterprise sector. Leading names here include GoDaddy Web Hosting, Intermedia Exchange, and Zoho Mail.

Finally, I offer a brief overview of the email marketing sector before considering it in more detail below. For many in the industry the term "email service provider" refers very specifically to companies that are involved in advertising and marketing via email. In their 2018 report, "State of Email Service Providers," the analytics company Litmus Software identifies the top US email providers as Mailchimp, Salesforce, Oracle, Constant Contact, Campaign Monitor, and HubSpot based on a study of nearly 3,000 marketers worldwide.[14] Mailchimp, made famous through its sponsorship of the widely lauded 2014 NPR *Serial* podcast, employees 800 staff, is estimated to be worth $525 million in annual revenue, and boasts a user base of 10 million people worldwide.[15] Mailchimp and its parent company Rocket Science Group make much of its independent financial structure and the fact it is still owned by its original founders and has resisted venture capital investment, in one case apparently rejecting a $1 billion offer.[16] In contrast, companies like Constant Contact and Campaign Monitor operate under various third-party investment structures including substantial venture capital. Constant Contact was listed as a public company in 2007, and in 2015 Endurance International bought the company for $1 billion.[17] As is often an unfortunate feature of acquisitions, Constant Contact sacked 15 percent of its staff two days after the sale.[18] Although the sector is served by over 200 different email marketing providers, it is becoming increasingly consolidated with the top 10 providers controlling nearly 63 percent of market share. Since 2013, there have been approximately $10 billion dollars in mergers and acquisitions within the industry.[19]

Email Software Programs or "Clients"

Many of the technical standards established in the 1970s and 1980s remain visible in the architecture of contemporary email programs. In fact, the scale, diversity, and functionality within the email client sector can be attributed to decisions of the period emphasizing that email readers should be flexible. "In contrast to paper-based communication," wrote the authors of RFC 733, the receiver "can exercise an extraordinary amount of control over the message's appearance."[20] Yet the eventual rollout of the email client industry was not inevitable. As Thomas Haigh has argued, a number of technical, economic and social forces combined to hamper the widespread adoption of very early email client programs. Haigh describes internet email software as "the business that wasn't" for two reasons. Partly the commercial opportunities didn't take hold in the late 1980s because Microsoft managed to kill the competition by giving away its own versions of email software. Additionally, the prevailing norms of the period emphasized open source programs written by amateur computer enthusiasts not looking to earn royalties, an environment that shaped the expectations of the market: "people were used to downloading programs rather than going to buy then in a store."[21] With the invention of the World Wide Web in the 1990s, however, web mail provision did start to become a viable business. Again, this was based on a "freemium" business model, with free email supplied initially but then monetized by inducing consumers to upgrade to versions offering greater storage or the capacity to download messages to local drives.[22] The email industry of the 1990s was populated by the enduring brands of Yahoo! and Outlook along with others that have since disappeared, many bought out by the dominant tech companies. In a 1997 guide called *Effective E-Mail: Clearly Explained*, Bradley Shimmin suggested the best email packages to try, including Pegasus, Banyan, Microsoft Outlook, Lotus Notes Mail, and Novell GroupWise. The guide itself came with a CD-ROM that offered readers a free version of Eudora Light to install.[23]

Today, email clients—or the mail user agents—operate along a number of axes: proprietary versus open source; browser, cloud- or webmail-based versus local, usually richer, client applications; and mobile device–enabled as against desktop accessed. In each of these scenarios, the protocols may differ; for example, Microsoft Outlook uses a proprietary version of SMTP and webmail generally handles email on the server itself while local clients allow download to a computer hard drive. But despite the varied functionalities and diversity of access to these email programs, the actual user interface has changed little in appearance during the last 40 years.[24] An obvious point perhaps, but the email client remains dominated by the memo template and postal iconography of envelopes and mailboxes.

So how do these images come to inhabit our desktops and mobiles? Which email brands define the visual landscape of our everyday social practices? One way to answer these questions is to chart the interconnected world of email client providers where the interface, its algorithms, operating system, and connectivity have a material impact upon commercial and personal use. As mentioned, this interdependency is apparent in the dispersed and variable ways the sector is technically and economically arranged.

While an ISP will offer email hosting, for instance, this might not be core to its business when seen at the level of support and functionality provided to email customers. Here we could include British Telecom, AOL, or iiNet, whose webmail applications are generally somewhat rudimentary. Occupying an opposite spot in the terrain sits the dedicated retail market of client software—including Microsoft Outlook, Apple Mail, Novell GroupWise, and Windows Mail—in which the interface is rich and will bundle calendar, editing tools, complex archival or search capacity, and encryption features. Sharing borders with both these categories are webmail, browser providers like Gmail, Yahoo! Mail, Outlook.com, Roundcube, and Fastmail. Here we see significant differences emerge that point to defining aspects of the email ecology system. While all are web based and generally free, these services differ regarding their software licensing arrangements, with Roundcube and Fastmail joining other open source providers like Mozilla Thunderbird to contrast sharply with Gmail's more overtly proprietary approach. Despite having released its application programming interface (API) to web mail developers in 2014, Gmail remains relatively closed to allowing fundamental modifications to its product.[25]

Case Study: Fastmail

In order to bring some specificity to the broad sketch of previous sections, I offer a brief case study of an email service provider. Fastmail, established in 1999 by Rob Mueller and Jeremy Howard, is an email hosting service based in Melbourne, Australia. It boasts an international customer base with more than 150,000 clients, both individuals and businesses, across 150 countries worldwide. In 2010 the software company Opera bought Fastmail, but in 2013 Opera discontinued their email client, Opera Mail, and Fastmail was sold back to its original investors. With a US office located in Philadelphia, Fastmail currently employs approximately 30 staff, and in 2015 it was able to acquire the email services US-based company IC Group with its products Pobox and Listbox.[26]

Fastmail suggested itself to me for a case study because it represents the diversity of the sector and offers the potential to understand the industry firsthand by speaking with one of its founders, Rob Mueller, and its chief of staff, Nicola Nye. I was interested to chart the business of email by examining its processes and systems of monetization to grasp its material and symbolic role in our everyday lives. I was also keen to explore how email itself functions as a workplace; we are all familiar with email as a key mode of communication within an organization, the topic in fact of the following chapter.

But perhaps what goes less remarked is what it's like to work at an email company. In terms of its business model, in many ways Fastmail is something of a pioneer in the email industry. One of its key objectives is to continue offering an advertising-free service to clients. The success of Fastmail has often been attributed to its abilities to resist advertising revenue usually generated through display ads and to avoid mining its users' data, a practice almost synonymous with the contextual advertising of Gmail. Instead users pay for the service, and although this is now a model with which we are familiar, when it launched this was actually new to the sector. From previous canvassing, we've seen that paying for specific online features, including an advertising-free environment, was not a common feature of the early internet landscape. But, back in the late 1990s, Mueller and his cofounder realized:

> There's a market for a professional service, one with no ads that's a fast interface for people who want email accessible anywhere and are willing to pay a bit for it. Which was pretty novel in 1999 saying "People will pay for internet services" because we're talking about the first dot com boom just as it was ending where everything was free. It was all about free free free.... And then we always said from the start... this will be a premium service and it's going to be charged. Which was novel in itself. But by 2002 we were actually charging people, and that's how we started. And we built what we thought was a better product basically, but one designed around the users rather than advertising revenue.

Nicola Nye agrees that "needing to put advertisers before consumers [is] only good for the advertisers and terrible for everyone else." What also differentiates Fastmail from others in the email provider sector is its attitude to scaling up and company development. Right from the beginning Fastmail has strongly resisted venture capital investment. As Mueller explains, "It was pretty rare at the time.... We took no external funding ever. We built it ourselves from scratch and we built it up bit by bit by bit." For Mueller this attitude has been a central aspect of the company profile:

> A thing we never really did was try to grow too fast, and we never had expectations of "we're going to explode overnight into this giant thing." We took it step by step and evenly, and we grew at a comfortable steady rate... building a small but sustainable business that doesn't take external funding, grows itself organically, which means that we're still in control of the business. We've been able to build a great product and we're continuously reaping the benefits of what we've built, and

> we're not subject to some arbitrary VC's growth rate requirements.... We're subject to what we want to do ourselves and what we think is the future and what we want to build out of our product which is a great place to be as a business.

In terms of the wider email industry, I was curious to learn whether Mueller's experience chimed with the annual "State of Email" reports from Litmus or others like them. I discovered that he too found the sector to be consolidating. Describing the situation from the end user perspective, Mueller explains:

> Anyone can be an email provider. Like I as an individual could set up a server, run email software on it, connect it to the internet and in theory I could send and receive email to that server. The problem is maintaining that as an individual these days is a lot of work. I need to set up all the software to handle spam filtering and antivirus and malware. And I want a way to access that email, so I need to set up a webmail system or something like that.... That's a lot of effort, you know? I would rather just outsource all that work to someone else. And these days you can get GSuite, what used to be Google Apps, Office 365, us, and a number of other players [for] all around the same—between $2 and $5 a month—and they will run the whole email service for you.

I also spoke with Mueller about my suggestion that we should identify an "email industry." Although he generally agreed that this mapping exercise works, he wasn't sure there would be much overlap between some of the categories. Specifically, he emphasized that marketing companies are a very different provider type than the hosting services of Fastmail or Gmail might be—in large part because of the sometimes questionable practices that certain email marketing companies can use to bypass spam filters and, more recently, to avoid the "Promotions Tab" in Gmail. Other strategies regularly employed are obscuring "unsubscribe links" in marketing material or purchasing email addresses from a third party. Mueller calls these techniques "grey hat marketing," a reference to hacker culture where the actions are somewhere on a continuum between malicious, intentional hacking (black) and hacking designed actually to help security or trust in a particular application (white). For Mueller, trust and reputation are key for the way that marketers conduct themselves within the sector, something he sees as inconsiderate of the end user. As he explains of email marketers and those who harvest email addresses for spam purposes:

> The biggest problem is that these guys on-sell to people who think "my email is the most important email in the world and every user wants it at the top of their inbox, can you get it in their inbox pinned at the top permanently in bold?"

Here Mueller is referring to situations where marketers routinely ignore consumer wishes. For example, if a marketing company is reported as spam, Mueller says they will often continue to send email and sometimes on-sell the address lists:

> Users can decide when they want their email, where it should go, how important it is to them. Because if they report it as spam, there should be a way to get off that list for good. But these guys hate this: "They said they wanted to sign up once ten years ago. You can never ever leave."

Fastmail is active in the open source community contributing to the development of projects including the Cyrus mail platform, a suite of programs that grew out of the Andrew Messaging System at Carnegie Mellon University. Similar Fastmail projects include leadership of the JMAP nonproprietary open protocol, a new standard for mail transfer aimed at creating "a bulwark against the growth of walled gardens."[27] Fastmail is a member of the advocacy group Electronic Frontiers Australia and makes regular submissions to the Australian government on security, privacy, and technology issues. It is a sponsor of "Write the Docs," an Australian initiative aimed at improving documentation practices in the software industry.[28] As Nye explains:

> Fastmail has developed, led and championed JMAP: the latest standard for email. . . . There haven't been systemic improvements in email technology in a long time. Individual corporations have been bringing out additional features but they have usually been confined to their . . . own customers.

Instead, she suggests, Fastmail has a significant public role to play because by "bringing out a new standard, it allows for everyone to build great new experiences and elevate the entire playing field."

Initially Fastmail provided free email accounts along similar lines to other browser-based email brands. During its affiliation with Opera Software, it scaled back its free accounts, moving to a paid service that provided larger mailbox storage and increased attachment limits. The current pricing structure at Fastmail offers plans from "Basic" costing $30 per year to the "Professional" account at $90 per year with file storage capacity, domain use, and archiving functionality differentiating the various products. To give some context, this is comparable with other paid browser-based email providers like GoDaddy, with yearly pricing plans ranging from $60 to $180; and Hushmail, also commercial free, which provides accounts for between $70 and $95 per year. Shifting from a free email provider to a paid account structure did affect its user base. Fastmail dropped from having millions of customers to numbers in the hundreds of thousands, a loss the company owners were prepared to incur because of their commitment to a commercial-free email interface environment.

At the time Fastmail was bought by Opera it was a small company employing only three full time staff who were focused on software development rather than increasing market share, self-confessed geeks who preferred to build new tools rather than create a sales and marketing department. As I mentioned earlier in this case study, "independence" seems always to have been at stake for the business, as its customers expressed fears when the sale to Opera was announced that Fastmail would become bloated and corporate. As one user posted at the time "Opera will probably ruin Fastmail by trying to turn it into just another Hotmail/Yahoo! Mail/Gmail competitor."[29] When Fastmail was sold back to its founders, independence was again a crucial part of the story. As Rob Mueller explained through the blog: "Developers and staff of Fastmail have now bought back the company. This means that Fastmail is once again an independent company, dedicated to building the best possible email experience for our users."[30] The technology news website ZDnet endorsed the sale because the "large email providers have been increasingly regarded with mistrust and scepticism."[31]

A further impact on the Fastmail brand is its domain-name transition. When it first launched, the domain Fastmail.com was already taken and so it registered as Fastmail.FM, a top-level domain extension associated with the Federated States of Micronesia and intended to draw attention to the business name through the use of its initials. This decision was made, as Mueller explains, in a fairly ad hoc manner with little thought to its branding implications. Over the intervening years, the slightly inexplicable extension had caused difficulties for asserting the brand within the email provider market and in 2012, as part of a re-profiling exercise sparked in part by the lead up to its split from Opera, Fastmail acquired the dot com extension from its owners, the job-search company Monster. For Fastmail, the dot-com domain would confer greater cache. As Mueller put it, "a short, sharp .COM domain is almost a luxury brand description, so it is actually signalling ... we are serious."[32]

In answer to some of my questions about working at an email company and the particular staff profiles or skill sets, Nye explains that until 2014

> Fastmail had less than 10 employees, was solely based in Melbourne, with the team all being developers with one front-end expert to craft our user interface. In 2015, we scaled up, acquiring IC Group (who made Pobox and Listbox) which brought to us another handful of people in Philadelphia, USA. We also hired dedicated QA, documentation, marketing staff and design staff in Melbourne. At this point we had around 15 people in Melbourne and 5 in Philadelphia.

At the time of writing, Fastmail employs 19 staff in Melbourne and 14 in Philadelphia. Roles include "operations staff, server-side and front-end developers, customer support, admin, design and mobile app developers." Working at Fastmail means being involved in the wider sector of email software development and provision. Staff members are encouraged to improve their skills and are given conference attendance budgets to keep up to date with advances in the field.

Having sketched the territory of email providers with a broad sweep of its constituent sites, and then by drilling down to explore an email company in more depth, I now look at impact measurements applied by the sector. And one useful way to do this is through email marketing, since here is where many of these tools are developed and deployed.

Email Marketing and Analytics

Email marketing is broadly defined as promoting goods and services through advertising campaigns whose mode of reaching audiences is via email communication. Third-party email marketing companies often handle these promotional strategies, and I discussed some of them earlier. The email-marketing sector is further bifurcated by those companies who test and analyze the *impact* of email rather than provide a mail delivery system. The two market leaders in email analytics are Litmus and Email on Acid; these services will check email content compatibility across different clients and devices, and supply marketing analytics that determine the rate at which emails are opened.

Opens—along with other metrics like bounces, clickthrough rates, deliverability, forwards, unsubscribes, and abuse complaint rates—help the email marketing industry to map campaigns across multiple platforms, devices, and software. As such, this generates a dense dataset through which to understand the material cultures of email and how its interface design and mode of access shapes user experience. Litmus, for example, regularly tracks annual open rates to determine market share in the email client sector with its 2019 calculations based on 10 billion opens worldwide.[33] Comparing how its client market share tracks over a number of years reveals the ebbs and flows of various email clients. So in 2011 Microsoft Outlook was the most popular email program accounting for 37 percent of all email opens, with Gmail lagging at only 4 percent. In 2012 Apple iPhone narrowly overtook Outlook for the no. 1 position at 20 percent of market share, with Outlook dropping to 18 percent and Gmail increasing to 5 percent. Retaining the top spot as most popular email provider and platform in 2013, Apple iPhone is followed by Microsoft Outlook, Google Android, Apple iPad, and Apple Mail. Sitting just outside the top 5 email providers is Gmail.[34] In 2015 these positions shifted and Gmail was ranked at second position in frequency of opens with 15 percent of market share compared to Apple iPhone at 32 percent.[35] By 2019 Gmail topped the leader board representing 29 percent of opens, Apple iPhone was a narrow second place with 27 percent, and Outlook sat at third place counting for 10 percent of email opens.[36]

Mobile traffic represents a significant shift in how we access our emails. Only 8 percent of email was opened on a mobile device in 2011 with the desktop occupying the top position. In 2019 these figures reversed and 42 percent of emails were accessed through a smartphone or tablet and only 18 percent of email opened on a desktop.[37] An increasing reliance on mobile email is particularly prevalent in the UK: a 2018 study found it had the highest mobile emailing rate worldwide, with 63 percent of respondents preferring to access email on a mobile device, almost double that of Latin America and the Caribbean, which sat at about 31 percent.[38] Reflecting the complex movement between mobiles and desktops, one study reported that a growing trend sees people using devices to "sort through" messages in the first instance before using a desktop later to reply. This practice seems to figure more highly in younger demographics with 40 percent of people aged 14 to 18 always reading emails on mobiles first and only 8 percent of those aged 56 to 67 adopting this kind of filtering.[39] Not surprisingly, many argue that changes to the platform of email access actually transform the message content. In a study of European office workers, more than a third of respondents felt that emails were getting less formal in style and messages were becoming briefer.[40]

My own surveys also delved into the range of email programs, providers, and devices constituting the email landscape in the United States and Australia. Microsoft Outlook and Gmail ranked highest in the types of programs people use, followed by Hotmail (owned by Microsoft since 1997) and Yahoo! Mail as strong contenders. Laptops, clearly ahead of desktops, represented the device most popular when accessing email. But the options of my survey further categorized mobile email use into iPhone, Android, Blackberry, and Windows Mobile, which when taken together support the worldwide trend already mentioned: reading emails on mobile devices is becoming the dominant mode of access.

Emoji

If our emails are becoming more informal and mobile, the rise of the emoji is pivotal to understanding these transformations. Designed by Shigetaka Kurita for implementation on the telecommunications provider NTT DoCoMo in the late 1990s, the term "emoji" combines the Japanese words for picture and character.[41] As a linguistic form, the emoji populates our communication landscapes across multiple devices and platforms including social media, websites, smart phones, and, of course, email. From a piece of pizza or phases of the moon to praying hands and the notorious eggplant, emoji seem to delight us. For the email industry, emoji are a significant form of communication and commerce. Campaign Monitor reports that emoji use in subject lines can substantially increase open rates.[42]

In the literature, there's often speculation about the differences between the emoji and that other 1990s' email convention known as the emoticon. The invention of the emoticon has been variously attributed. Most accounts credit Scott Fahlman at Carnegie Mellon in 1982 with the first internet appearance of an emoticon smiley face, while others claim that Kevin MacKenzie established the genre in 1979.[43] Such ambiguity over the precise invention is not remarkable considering the long trajectory of nonverbal language representation within written texts, where for centuries faces and emotions have been expressed through punctuation and other symbols.[44] Although similar in their visual composition and in their function as supplements to written text, emoticons and emoji differ technically and semantically. Emoticons are usually restricted to execution through the alphanumeric typographical keys within text-only environments and in general denote facial expressions and body language. Meanwhile, emoji have a far wider range of meaning than emoticons since they can be used to describe ideas, situations, feelings, animals, objects, food, weather conditions, and places. Emoji are also visually richer because they are rendered as images through Unicode character sets. One way to think about this technical difference is to recognize that emoji often rely on proprietary software so that not all browsers, operating systems, or devices will successfully translate a given image.

Despite issues of interoperability, however, emoji have reached near ubiquitous status worldwide as Apple and Google launch ever more niche symbols to convey meaning by visually enacting humor, irony, or pathos. The "Face with Tears of Joy" emoji was selected by the Oxford Dictionary as its Word of the Year in 2015. Reflecting its substantial social impact, this was the first time an icon had been chosen by the dictionary; the decision was based on frequency and usage statistics across a range of emoji appearing worldwide.[45] As sites for expressing cultural identity, the emoji repertoire has regularly introduced images that speak to issues of diversity and inclusion. Expressing growing unease about the homogenous nature of the available icon archive, several initiatives have urged designers and commercial providers to more fully embody the diversity of the emoji user community. As the campaign "Diversify My Emoji" exclaimed in 2014, "there are currently over 800 emoji, but only two represent people of color."[46] Responding to such concerns, in 2015 the Unicode Consortium, the industry body responsible for computer language standardization, launched new skin tone symbols for human emoji based on the Fitzpatrick scale of dermatology, because, as they put it, "people all over the world want to have emoji that reflect more human diversity, especially for skin tone."[47] Following suit, the commercial sector soon embraced the push for greater racial diversity. In June 2016 Facebook Messenger introduced 1,500 new emoji, promising that "Messenger is beginning to make emoji more representative of the world we live in … [from] skin tones that you can choose to lots of women in great roles."[48]

Given the ubiquitous presence of emoji rich with meaning, it's not surprising that questions of their diversity continue to press the public consciousness. To what degree emoji representation either bolsters or ameliorates racial inequality is a topic that continues to prompt debate.[49] Emoji also offer impetus for public awareness campaigns to target bullying. The first Unicode-approved social cause symbol was rolled out in 2015 by a campaign titled "I Am a Witness" to combat abusive online behavior: appearing as an eye inside a speech bubble to illustrate the notion of bearing witness, this emoji was available on iOS 9.1 and Android 6.0.1. The idea behind the initiative was that people could help moderate and govern their social media spaces to configure safer environments. The organizers outlined their objective by exclaiming, "You've seen it: someone leaves a rude comment on someone's page or video. Use the emoji to show it's not okay."[50]

Against such efforts of inclusion and diversity, however, many critics point to the ways in which emoji may in fact shut down creative expression to become, in the words of Luke Stark and Kate Crawford, a form of conservatism. Here, emoji function as affective labor within market logics. As they elaborate:

> Emoji offer us more than just a cute way of "humanizing" the platforms we inhabit: they also remind us of how informational capital continually seeks to instrumentalize, analyze, monetize, and standardize affect.... Emoji are an exuberant form of social expression but they are also just another means to lure consumers to a platform, to extract data from them more efficiently, and to express a normative, consumerist, and predominantly cheery world-view.[51]

Umashanthi Pavalanathan and Jacob Eisenstein offer a similar perspective. For these authors, the different historical trajectories of emoticons and emoji are crucial to understanding transformations to digital language. They argue that the two icons compete for social media attention. When people deploy emoji in a tweet, for example, the frequency of their use decreases of use of emoticons and other nonstandard forms of lengthening and capitalization (using "SOOOO" instead of "so"). The increasing dominance of emoji over other paralinguistic expressions represents a unique moment because unlike emoticons that have evolved organically from users themselves, emoji are the products of industry and commercial bodies. As they put it:

> The stakes of this competition are high because emojis differ from other nonstandard forms in one crucial way: unlike emoticons, phonetic spellings, and non-standard orthographies, emojis are not the result of individual innovations propagating through a social network. Rather, they are created and adopted through a standardization process that is run by... large software and technology companies. In sociolinguistic terms, emojis are "change from above," while other paralinguistic cues are "change from below."[52]

As these studies make clear, emoji are a particularly salient form of communication for the marketing industry. And a number of projects have attempted to gauge emotional content, with the first emoji-sentiment ranking published in December 2015 by Petra Kralj Novak and colleagues.[53] To arrive at their lexicon the authors analyzed 1.6 million tweets in 13 European languages during a two-year period and ranked them according to the sentiment classification "positive, neutral, or negative." Contrasting the treatment of emoji with emoticons, the study found that Twitter users tend to locate emoticons at the end of their tweets but emoji are peppered liberally throughout a given post. Where emoticons are used singularly, emoji appear in clusters. Drilling down further, the study reveals an interesting practice: people place their highly affective emoji, whether positive or negative, toward the end of their tweets. Neutral emoji, for example musical notes or the copyright sign, occur toward the beginning of a post. In general, the authors found that the most highly used negative emoji consist of faces, whereas positive emoji were indicated more broadly by faces and also hearts, party symbols, wrapped presents, and trophies. If some of the results seem self-evident, certain findings—in particular the police officer emoji ranking as negative—point to the rich social data pulsing through these innocuous images. No wonder the email marketing and advertising sector is increasingly drawn to emoji analytics.

For email marketing providers like Mailchimp, the importance of emoji has increased with the growing reliance on mobile screens: less viewing estate demands more semantic precision. Tracking how their customers use emoji in email subject lines, Mailchimp results differ slightly from the major surveys mentioned above. The "Face with Tears of Joy" is nowhere to be found on the list of most popular campaign emoji. Instead, the registered symbol, "R," ranks as the most often sent image, with the big-eyed happy emoji and the smiley with heart eyes in second and third place respectively. Reflecting its commercial customer base, the flight or airplane emoji also figured highly on their frequency rankings. During a three-month period in 2015, Mailchimp found that 1.4 billion emoji were sent encompassing 214,000 campaigns. Mailchimp was also interested in the combination of emoji sent with 31 percent of advertisers sending multiple pictures. Visualizing a graph of the emoji network, Mailchimp plotted patterns and combinations revealing that similarity prevails. Food and beverage pictures were sent together, likewise shirts, ties, and coats were all grouped in the same email subject line. More surprising, perhaps, was to find the applause icon nestling with the Edward Munch inspired "Face Screaming in Fear."[54]

If you're worried that I've fallen too far into banality, let me remind you that this book is about the everyday contexts for email. What could seem like excessive attention paid to the tiny calibrations of emoji is for me entirely germane to the wider project. In this regard, it is important to note the considerable technical problems that emerge with interoperability and the inconsistent rendering of emoji. Earlier I mentioned in passing that different operating systems, devices, and email clients offer varying levels of support for emoji. Browser rendering, for example, is notoriously uneven with the dreaded empty box appearing regularly in place of a desired picture. This fact is often overlooked in much of the academic literature so that a kind of universality prevails where it is assumed we are all dealing with images; this assumption is strengthened by the claim that emoji represents a new form of communication with the capacity to transcend barriers of national languages.[55] Here is where, again, the commercial sector of email is leading the analytics research with email marketing companies pointing out the folly in expecting emoji to be accurately rendered. If you use emoji in place of crucial words in the subject line about a product or service, you risk alienating your customers, they warn. As Litmus puts it, "Nothing says 'unsubscribe' faster than incoherent emails.… If emojis don't render? You're basically sending an email with a blank subject line."[56]

Focusing on the varying patterns and success of emoji rendering also helps enhance our map of the email provider sector. We can see, for example, how emoji use impacts mobile email traffic. Mailchimp reports those businesses who target people with high levels of smartphone integration will tend to adopt emoji in their campaigns more frequently than those whose subscribers rely on web or desktop email clients. Or put another way, if you are marketing to someone who opens your email using Outlook or Hotmail, it's not a good idea to include an emoji. Outlook recipients, for instance, open only 2 percent of messages from businesses that use emoji in their campaigns.[57] Adding specificity to the calculation of emoji translation, email marketing companies will often supply resources that help test and visualize the way a campaign will be received. This is conducted through "subject line checkers" or "A/B testing," where slight variations are made to a given campaign in order to evaluate open rates. Does including a recipient's name in the subject line increase success? And, relevant here, how do emoji appear across a range of email software clients? As Mailchimp explains, emoji support is significantly unreliable even from within the same email client, with Gmail that is accessed via Desktop Chrome showing different emoji than the same Gmail message opened with Mobile

Chrome.[58] Tracking the path of an email, the time and place it was read, or whether its links were followed, tells us a great deal about how people use email in everyday life.

Before leaving this section it's worth noting the industrial connections between the beginnings of emoji use in Japan and the development of mobile email. In the United States, United Kingdom, France, Norway, and Australia, email and SMS are viewed technologically as quite distinct communication platforms. But in Japan email and texting are often seen as almost synonymous. This is because historically the SMS protocol has faced quite challenging problems with standardization, which means its use is regularly limited to message exchange within the same telecommunications provider.

Although the situation is getting easier, for many years it was not possible to SMS someone outside of your own mobile carrier.[59] Email, however, offers vendor-neutral interaction and has historically dominated the mobile messaging market. Such technical barriers to SMS uptake explain, in part, the contrasting usage patterns across different geographical markets. A 2010 study of US, European, and Japanese mobile phone behavior found that only 40 percent of those surveyed sent mobile text messages contrasted with nearly 67 percent of their US counterparts and 81 percent of Europeans. Conversely, the figures for mobile email use were much higher in Japan.[60] Where this survey was able to distinguish between SMS and email use, other studies of Japanese mobile phone consumption have argued that SMS and email should be subsumed under the category of "mail," and then the focus is mobility of messages and senders rather than the particular communication protocols. Indeed, as Kenichi Ishii explains from the user perspective, email and SMS in Japan often converge so that consumers do not really distinguish between the two services.[61] In their research on Japanese phone usage, Mizuko Ito and Daisuke Okabe define mobile email or *keitai email* as encompassing "all types of textual and pictorial transmission via mobile phones," and that "includes what Japanese refer to as 'short mail' or 'short messages' and Europeans refer to as 'short text messages,' as well as the wider variety of email communications enabled by the keitai Internet."[62]

For me, the real story of these intercultural comparisons between email and SMS is how early Japan adopted mobile email. Moreover, the uptake of *keitai* email actually drove the success of the mobile data service industry, sparking new revenue streams for infrastructure and content providers. Just as it had accomplished with the dramatic spread of the internet a decade earlier, in 2000 email was once again being hailed as innovative. This time it was for its success across NTT DoCoMo's i-mode, one of Japan's leading mobile internet carriers and the birthplace of emoji, and for its capacity to generate new users and traffic. Although game downloads and internet browsing accounted for a significant portion of subscriber activity within the network, it was email that was attracting the customers. In 2001 nearly 60 percent of mobile traffic was produced by email, and by 2005 that figure had increased to 83 percent.[63] In 2006, research was already speculating whether *keitai* email would soon be replacing PC-based email.[64]

To provide some context, remember that figures for US adoption of mobile email reveals that less than 10 percent of email traffic in 2011 occurred on mobile devices. Enthusiasm for such figures does need to be tempered somewhat by the acknowledgement that mobile internet, rather than PC-enabled internet, was more common earlier in locations like Japan and China than in the West.[65] But as a number of scholars have noted, it was the rapid diffusion of i-mode in Japan that helped set the industry standard for mobile internet worldwide, pushing other markets to adopt the devices and software necessary for its diffusion.[66]

The relevance of emoji here is that its use in emails was instrumental to the roll out of the Apple iPhone in Japan. Matt Alt argues that, unlike most of the other markets worldwide, when the iPhone launched in 2008 it was not an immediate success in the Japanese telecommunications market. What prevented its take-up was that initially it did not support emoji use because US consumers had not demanded emoji. As it became obvious that substantial numbers of potential customers were rejecting the iPhone, Masayoshi Son, then the president of Softbank, Apple's Japanese retailer, announced at a press conference he had successfully convinced Apple "that email without emoji isn't email in Japan."[67]

Behind the Analytics

Though the figures discussed earlier offer an evocative snapshot of the rise and fall of major email brands and their platforms, the reporting algorithms themselves tell an equally compelling story. Canny readers, even the mildest of media materialists, will have spotted a flaw in some of the email analytics discussed earlier. You might recall that we drew on "top 10 email client lists," figures that show, for example, how iPhone gets top billing as most popular email tool beating Outlook, or that iPad email might increase its market share in a given survey period. But comparing Outlook to an iPhone doesn't really hold since the former is an email client—a program that sits on hardware—whereas the iPhone and iPad are devices and, taking it further, Google Android is an operating system. Drilling down to discover just how this data is calculated reveals, once again, the intertwining socio-technical forces animating the email ecology system.

One of the reasons for the dominance of the iPhone in the Litmus reports is that, by default, opens detected on a mobile device are classified as mobile opens irrespective of the particular application used. So if a consumer accesses email through a third-party application like Gmail or Yahoo! on a mobile rather than use the native client (Mail in the case of iPhone) this will be reported as a mobile open by Litmus to its customers. Similarly, webmail-based programs opened on a smart phone register as mobile access rather than count as the particular web client. It's important to point out that this is not a decision made only at the marketing statistics level but is often built into the software applications themselves. As Litmus explains:

> Various mobile applications may report differently, depending on how the application was built. For example, opens made in the AOL application for iOS/iPad will report as "AOL Mail using iPad" within your Email Analytics report, whereas the Yahoo! Mail application for iPhone reports as "Apple iPhone."[68]

In other words, if you are trying to get information about the popularity of, say, the Yahoo! Mail client, the results will be unreliable because it collapses under the larger category of Apple iPhone. The software for downloading images and managing inboxes also has an impact on the collection of data. This is because the central tool used for collecting metrics about email opens is a tiny 1 × 1 pixel-sized image embedded by email marketing programs into their advertising. When an email reaches its recipient the image is downloaded, and information is then retrieved about where and how the image is accessed.

Gmail is an illustrative case, and accounting for the way it shifts positions in the rankings over time are the various updates and new functions added to its interface. In 2013 Gmail introduced its "tab" inbox system, which automatically filters incoming emails into the three categories of Primary, Social, and Promotions. Not surprisingly, email marketing companies feared this would have a negative effect on the rates by which people would encounter advertising messages because any mail containing "unsubscribe" links or other advertising identifying features would bypass the primary mailbox. In response, and remembering the earlier comments from Fastmail's Rob Mueller, Mailchimp conducted a study to compare rates before and after the new tabbed inbox was introduced. Examining 1.5 billion emails for a six-week period it found a noticeable drop in opens, sparking discussions about what workarounds might ensure advertisers can "get out of the Promotions tab and into the Primary tab."[69] Although it tested various software solutions Mailchimp concluded that regrettably it had "no control over the placement of emails in Gmail, and there is no proven way to 'beat' Gmail's algorithms." Instead, Mailchimp recommended that users engage their customers to a sufficient level that they themselves would move the marketing messages to their own primary tab.[70] In addition, as part of its email testing service Litmus presents an interface that helps marketers determine "which Gmail tab will your email appear under?"[71]

During 2013, a series of updates governing how images download in Gmail were also implemented affecting the collection and reporting of data about opens. Where previously Gmail, like most email clients,[72] would link to images stored on third-party servers, the updates meant that Gmail began caching images on its own servers. The new Gmail configuration also automatically displayed embedded images within the message rather than prompting the user with the option "display images below." These changes make tracking email rates much more difficult to achieve because the referring data provided to a marketer—device, geolocation, or site accessed, for example—reports only as a Google caching server. So "an email opened in Gmail with a web browser will be indistinguishable from an email opened in a Gmail mobile app." Why would Gmail make these changes? Although Google explained that the changes improved functionality by providing a faster viewing experience, and would improve security by checking images for malware hosted on its own servers, it's clear the new design brings commercial advantage to Gmail. It makes business sense that Gmail would want to limit access by its advertising competitors to consumer data like IP address or referring location. Moreover Google itself relies on its own AdSense app for generating advertising revenue rather than use email campaigns. Making email marketing less attractive through its tabbed in box also seems logical.

Conclusions

Landscapes are easy to overlook. Because they form the backdrop of our daily journeys and everyday vision, we tend to let them play a secondary role. But landscapes can also suddenly slide into view, their contours becoming the main focus. In this chapter I've shown the rich mediascape of email service providers, calibrated by a multiplicity of software design decisions, varied commercial approaches, and different modes of user interface engagement. In order to bring the landscape to life, I sketched the broad elements of an email industry, its major players, company profiles, and market share. One of the first challenges in such a project, conceiving of email *as* an industry in its own right, is the irregular way that market value is calculated. Some reports will exclude the email marketing sector from their assessment while others will look only at business email rather than at the consumer or personal user. Despite the different approaches to segmentation it is possible to make some broad statements about the value of the sector and look at its predictive growth. The email industry, worth around $40 billion annually worldwide, is expected in the next decade to increasingly rely on cloud computing.

The lack of agreement about how to carve up or calculate market share in the sector is mirrored by the imprecise way that various terms are used. Those in the email hosting business see their role as "email service provider" specifically in reference to tailored email addresses, shared calendars, and contacts, strong encryption, spam filters, and mailbox search functions. But email marketing companies also use "email service provider" to refer to their function in the design and rollout of campaigns. Such ambiguity is not simply about what to call things but reflects fundamental disagreements about how people actually want to use email in their everyday lives. This tension became clear to me when I talked with Rob Mueller, the director of the email service provider Fastmail, the subject of the brief case study I conducted for this chapter. Many of the strategies discussed above that marketers might use to get their emails opened—by avoiding the spam folder or disguising opt-out methods, for example—can be viewed either as effective communication or gaming the system, depending on where you sit in the industry.

Although Mueller was careful to point out that in many cases email marketing companies might think they are doing the right thing by their clients, in actual fact they need to listen more closely to end users who are often quite clear that they don't want to read a particular promotional email. Trying to hoodwink them into opening your advertising, "avoiding the promotions tab in Gmail," is expressly ignoring their wishes. Seen from the other side, however, while researching the social media channels, blogs, and conference sites of these email marketing providers, it's equally clear these are everyday people themselves going about their ordinary jobs in content creation, design, and sales. Watching numerous panel discussions, workshop presentations, and seminar videos I was struck with the obvious yet profound fact that since email represents a billion-dollar industry, it is a major source of job creation and employs thousands of people worldwide. Attempting to get a sense of how the industry functions it was important, therefore, to find out through the Fastmail case study the professional profile of staff, their skill sets and views of the industry.

Another way to understand the industry is to focus on the ebbs and flows of particular email brands. Which email clients still dominate the terrain and which have fallen by the wayside? Using metrics such as "opens," it is possible to glean information about the steady rise of Gmail, let's say, and the increasing shift from desktop to mobile email. This in turn helps to understand how people encounter email in their day-to-day lives by plotting the software and devices through which they read their messages. But as with all data analytics, often the methods used for gathering the information may skew the results. In certain cases this is because the email software itself embeds specific reporting statements (opened on Apple iPhone irrespective of email client, for example) or it rests with the statistical decisions made by analytics companies.

Beyond these granular-level constraints, however, it's important to note that email analytics itself is a business. Being mindful of how a particular report is funded is a necessary step in parsing the findings of many of these studies. This does not mean these figures about email use, such as how often people read emails or which software is most popular, are completely unreliable. But as a researcher it would be naive in the extreme not to notice that many of these industry reports have a vested interest in, frankly, promulgating the efficacy and impact of email as a marketing tool. No one is likely to read a report from such a source that says, "People aren't opening emails." Driven by an increasing frustration about the unavailability of independent figures for email use, in 2015 and 2017 I conducted two major surveys of email use, in Australia and the United States respectively. While in many ways my findings echoed those of the commercial endeavors, it was satisfying having gleaned these answers on a slightly more independent basis.

Bringing together many of these issues about the different email industry stakeholders, data collection, and consumer choice is the emoji. Here is where divergent views are keenly felt about the commercialization of our communication channels. For many in the email provider sector, emoji are evidence of how everyday life has been corporatized with prefabricated symbols of emotion designed only to sell product and to monetize our affective labor. For those in marketing, emoji are an obviously rich resource to enhance advertising and communication techniques. Because of this, scholars have pointed to the overweening sense of commercialization found in the emoji. When contrasted with the "grassroots" emergence of the emoticon from internet users themselves, the provenance of the emoji is decidedly commercial in its reliance on proprietary software. Certainly, my survey respondents were vocal about the place of emoji in contemporary email cultures. As evidence of its affective power, respondents often agreed that emoji had no role in the workplace because they conveyed too much or the wrong kind of emotion for a professional setting. I expand on these aspects of email practice in part II, "Affect and Labor," where I look at how bureaucratic intensity shapes the workplace.

II Affect and Labor

4 Bureaucratic Intensity and Email in the Workplace

Slack is the name of a cloud-based application for organizational collaboration that gained traction in 2015. Carrying the slogan "Where Work Happens," Slack's developers promised it would replace or even kill email entirely. One of its strengths is its apparent capacity to encourage team communication and to "subvert organizational hierarchy."[1] Whether Slack will eventually replace email communication by flattening work hierarchies is still unknown. What is certain, however, is that the exact same claim has been made before—about email!

During the late 1980s and into the mid-1990s the emerging field of CMC (computer mediated communication), together with the newly minted *Wired* generation of tech writers, regularly predicted that with email came easy access to your boss. The RAND Corporation published its *Ethics and Etiquette for Electronic Mail* in 1985, observing that a "junior executive can send a message to a senior executive, bypassing several levels of control. 'Electronic Mail' tends to be more democratic."[2] Similarly, in their 1986 research on the impact of email in organizational behavior, Lee Sproull and Sara Kiesler identified that "status equalization" occurred because in many cases senior management were more accessible to staff via email rather than in person.[3] This suggests that the desire for an ideal office operating beyond the limits of hierarchy is an enduring one and, further, that new technology is at its heart.

In chapter 1, I argued that the prehistory of email design in the memo template locates us materially and symbolically within the flows of bureaucratic life. Reading against those who dismiss bureaucratic structures for being stultifying, unproductive, and uncreative I tried to show instead the central place of bureaucracy in producing the software interface of email communication. In this chapter I want to extend the argument by focusing more explicitly on the role of email within the institution. Here, I am interested to tease out how the bureaucratic register is in fact lively and dynamic; how, contrary to popular belief, bureaucracies and their media forms actually move us in emotional ways and inflect our communication with affect.[4]

Email has become synonymous with the workplace, so I begin the chapter by identifying five main features that help define organizational use of email. I then map the reach and scope of email across a broad range of demographics to identify what we actually know about how different people use it at work. Evidence for this comes from a variety of qualitative and quantitative data including my own two surveys. Having charted the patterns of usage for workplace email I place this within the context of office life and bureaucratic management. In this chapter, therefore, I continue the material focus on email to understand it as a document of "organizational authority"[5] that is, nonetheless, experienced as routine and banal. Its banality again points us in the direction of what I am calling "bureaucratic intensity"—the ways in which our everyday media practices structure the rhythm of workplace communication, producing flashes of anger and irritation or moments of pleasure and stretches of boredom.

To bring texture and depth to the debate, I look briefly at the scholarship on bureaucratic knowledge production to understand its processes. In particular, it's important to get a sense of the different ways that the term "bureaucracy" is deployed. Much of this literature deals with the regulatory systems of the state investigating the institutions that underpin transnational networks of geopolitical diplomacy and international relations.[6] While this research is useful for glimpsing how people deal with the processes and discourses of governance, investigating the mechanisms and social relations of the World Bank or WTO, for example, the point of departure for this chapter is with the everyday administration of an organization, often called "red tape." It is, however, almost impossible to untangle the two conceptual threads of the term 'bureaucracy." What links state power and paperwork is an intensity of feeling expressed toward the discursive and material systems of management. As Ben Kafka so eloquently explains in his historical exploration of paperwork, the term "bureaucracy"

> became a vague expression of an even vaguer sense that something, or someone, or many someones—anyone or anything but the structural contradictions of the liberal-democratic project as such—had to be to blame for how much paperwork was required not only to govern, but to be governed in the modern world.[7]

I conclude the chapter with a discussion of the "reply all" disaster, surely one of the stickiest sites for workplace affect. People's perceptions about their email use are often bound up with how well they can negotiate workplace governance. I also revisit the emoji. From thinking about how it functions as a commercial product within email industry contexts in the previous chapter, I now chart how it figures as a form of workplace communication. Although email companies attempt to monetize the uptake of emoji, many people actually reject using them in professional contexts. More than 60 percent of respondents to my US survey did not think emoji were appropriate for workplace purposes. "Use your words," as one person put it somewhat caustically.

The Defining Features of Workplace Email

As I discussed in the introduction, much of the research conducted during the last 40 years about email focuses on its use in organizations and workplaces. In this section I briefly outline the major themes that have preoccupied the field together with some of the competing methodological and critical perspectives to these topics. Here I am not attempting a detailed literature review because that would completely dwarf the book itself, so extensive is the research on workplace email. Indeed, more than two decades ago Michael Holms warned, "No survey of e-mail research is likely to be complete or even temporarily so. The rapid diffusion of organizational e-mail has earned it vigorous popular and scholarly attention."[8] So instead I attempt to establish what we know about organizational email by identifying five broad topic categories that will highlight the key features of email use in professional contexts. While each area opens up to encompass a range of associated concerns, broadly speaking the aspects of email with which workplaces tend to grapple are: governance, productivity, monitoring, device, and affect.

Some of the earliest approaches fall under the category of governance (or structure). These include research already noted that looks at how email functions within a workplace either to collapse or entrench hierarchies. Surveying the early literature, Laura Garton and Barry Wellman argued that email use in organizations provided "more participatory and egalitarian decision making" because it reduced "status cues."[9] Writing in 1988 for *The Journal of Business Communication*, John Sherblom explored whether email signatures hold clues about attitudes to workplace management and found that organizational hierarchy can predict signature variation. When people send emails to their supervisors, they included lengthy signatures. Conversely managers tend to omit signatures when emailing "downward."[10] Further research has considered how the opening and closing salutations of emails change relative to company status;[11] how email can function as a "hidden weapon" of dispute resolution in labor relations;[12] whether certain phrases used in an email will signal a person's leadership role within a company;[13] and how the introduction of email into an organization enables "power games" to occur.[14]

Workplace email has also raised questions about productivity and the related areas of information overload, workflow interruptions, and work/life balance.[15] One of the thorny problems in the scholarly literature is whether a causal link can in fact be found between increased diffusion of email in the workplace and stress-related problems, decreases in productivity, or fragmented workflows. The difficulty, as Judy Wajcman explains, is particularly noticeable within assumptions about email and information overload. What she calls "the mechanistic approach" tends to view interruptions, from email or other ICTs, as necessarily negative and distracting from the real tasks at hand. Consequently, the broader picture of a multidirectional engagement with diverse types of media is overlooked. As she puts it, "Knowledge workers now inhabit an environment where ICTs are ubiquitous, presenting simultaneous, multiple and ever-present calls on their attention. These mediated interactions can no longer be only framed as sources of constant interruptions that fragment time."[16] In a sense, the sheer volume of literature about workplace productivity, from both popular and scholarly sources, animates Melissa Gregg's monograph study *Counterproductive: Time Management in the Knowledge Economy*. As the title suggests, here Gregg takes a step back to critique the very notion of productivity itself and, through historical and recent case studies, she questions the incessant pressure to "get things done" in our contemporary places of work. While email is not the only site of analysis—domestic labor-saving devices, mindfulness apps, and self-help manuals all help to fashion the productivity imperative—it does figure highly for Gregg. In particular, the email strategies of "inbox zero" may give to workers a momentary feeling of accomplishment that often "reflects an inability to influence the broader agenda governing one's work."[17]

Under the category of "monitoring," as we saw in the introduction, many workplaces and staff have worried about the increasing potential for surveillance presented by email. In addition to those studies already discussed on employee privacy rights, other researchers have considered new forms of encryption and the cost to industry of "data leakage," the inadvertent or intended circulation of sensitive company information.[18] In my own surveys of Australian and US email users, I found a range of opinions on workplace monitoring, as I elaborate below.

Where once it was safe to assume that emails were opened only on a desktop at work, increasingly the platform or device has assumed significance. One of the first studies to bring this to our attention, by Melissa Mazmanian, Wanda Orlikowski, and JoAnne Yates, looked at the impact of wireless email devices in the workplace and showed how information professionals "negotiate numerous and often conflicting organizational expectations, personal goals, and properties of a technology."[19] In a subsequent paper the authors coined the phrase the "autonomy paradox" to describe how knowledge workers using mobile email devices may initially feel liberated by being able to check email in different contexts, but this connectivity can actually increase anxiety and the "always on" work environment.[20]

Finally, workplace email is often viewed as a site for affect. From outbursts of anger through to the conciliatory use of an emoji, employees in organizational settings frequently wonder about the emotional dimensions of email. Is there a way systematically to measure affect and therefore to predict or regulate its appearance through the email channels of a workplace?[21] Not only can negative emotion impact upon an employee's immediate work environment it can "spill over" into home life.[22] As we saw earlier, much of this research is conducted through behaviorist or psychological frames that use personality scales to develop strategies that minimize workplace disruption.[23] I am interested in the affective register of email, but I want to take the inquiry in a slightly different critical direction than these studies. Instead of locating the agency entirely with the individual, I will look further afield to the organizational logics of the workplace. Easy to overlook and almost taken for granted, the everyday and routine practices of bureaucracy represent a useful way to look at how email operates in work life.

Having discussed the main issues that arise when considering workplace email under broad categories, I now zoom in to provide a snapshot of its use as seen through demographic figures showing uptake and reach.

Mapping Email Use in the Workplace

Despite Slack-like software snapping at its heels, email still dominates the workplace. A 2018 survey of US knowledge workers conducted by the cloud computing provider Intermedia found 87 percent of respondents ranked email as the primary mode of communication used by employees within their organization.[24] In 2016 a study of 9,000 knowledge workers in the United Kingdom, United States, and Germany reported that 74 percent of respondents cited email as the "most essential communication tool," with face-to-face conversations ranked second at 49 percent. The 2016 survey was conducted by Unify, a cloud communication software company that defines knowledge workers as those "whose main capital is knowledge, whose job is to 'think for a living' and who have access to technology as part of their day-today jobs." When asked which communication technologies they would like to see "removed completely from the workplace," 28 percent of the knowledge workers who responded answered it was email.[25]

Unfortunately, Unify did not publish further findings of the report. It would have been useful to see what other applications these knowledge workers wanted removed from the workplace. It is probably worth noting that Unify is owned by Atos—the company we met in the introduction who has become famous for imposing an email ban within its organization. One does not need to be a conspiracy theorist to make the simple observation that if you are producing organizational software aimed at building a new collaboration platform you might want to scale back the significance of email. In 2018 Unify ran another of its "The Way We Work" surveys. This time their focus was remote working and the ROWE framework. The acronym stands for Results Oriented Work Environments and sums up the principle that it shouldn't matter where employees carry out their work as long as the results or outcomes are achieved.

Discussing the 2018 Unify survey, one analyst commented that trying to achieve a so-called work/life balance is the wrong approach to take. Rather it is work-life *integration* that's needed. As he explains, "Work-life balance implies ... clear boundaries between work and life, whereas work-life integration acknowledges that boundaries no longer exist."[26] To me this is quite a chilling prophecy (perhaps akin to the death-of-privacy claim so beloved by social media companies). But, as I'll discuss, in the scholarly research there are efforts to add nuance to simplistic on/off binaries when thinking through the different domains we inhabit in our work lives.

Trying to capture the patterns of workplace email use often means interpreting studies whose main aim may lie elsewhere. The cloud security company Okta conducts an annual "Business at Work" survey of workplace technology drawing from its global customer base and analyzing millions of daily authentication and verification interactions. Okta's 2019 report of the most popular work-based applications shows that Microsoft Office 365, combining file management, word processing, and email, tops the list both in number of customers and active unique users. Sitting at fourth place in their list of the top 15 cloud-based applications is G Suite (including Gmail) followed by Slack, Zoom, and Dropbox among others.[27] Contrasting the uptake of social media and newer collaborative tools versus email, a survey of communications professionals in 2017 found email to be the most popular internal workplace channel with 95 percent of respondents reporting it as their preferred method. Although over 90 percent of those surveyed used email as their dominant mode for contacting people outside their organization, only 27 percent believed it was the most effective tool, citing social media as a potential alternative. Nevertheless, according to this study, email remains the "go-to method for business communication."[28]

My own surveys of email users in the United States and Australia support some of these figures. The results showed that email is second only to face-to-face communication as the most frequently used mode of communication at work, with 84 percent of Australians using email "often or quite often" compared to 85 percent for face-to-face. Similar figures emerged from the US study where again email was the second most frequently used form of workplace communication, only just edged out by face-to-face situations. For Australian workers, the corresponding figure for internal social network programs like Slack, Yammer, or Asana is 12.5 percent, which has yet to rival the use of fax at 16.2 percent. A quarter of employed Australians use social media "often or quite often" to communicate at work, while the phone remains popular at nearly 79 percent. Fewer than one in five workers said they used email less often than they did five years ago, and only three in ten respondents agreed that social media had replaced email in both work and personal situations. Figures like this could suggest a sharp distinction operating between workplace and personal communication, as if every email opened at work is for job-related purposes. I was interested in how people juggled their communication practices across contexts and devices; the survey asked how many accounts people used and whether they kept distinct social and work email addresses. While eight in ten Australian workers generally maintain separate accounts for work and personal use, nearly four in ten people admitted they send some personal emails from their work account.

Being curious to know whether people worried that these personal emails could become public and more generally what attitudes circulate about how the workplace should manage the blurring of work/life boundaries, I included survey questions about access. I found that people are divided when it comes to whether employers should have the right to access email accounts maintained at work. More than half of respondents felt workplaces should not be able to monitor email, whereas others reasoned access was justified as long as the content was work related or posed issues of organizational liability. One survey participant explains this point:

> If it's work email address, then yes, they should. You should not use your work email address for anything other than work related use. You are there to work, for which you are paid. So work and use your personal time for personal things such as sending emails.

If workplace email monitoring has been an area of substantial research over the last three decades, then what about the flipside? While companies may fret about revenue and labor lost to personal or social activity in the workplace, let's remember all those emails checked, answered, or filed while not at work. Keen to expand the narratives of the "always on" workforce I discussed in the introduction, my surveys investigated the interplay between email and work/life balance. More than a quarter of survey respondents in Australia said their workplace expected them to be contactable via email outside of normal working hours.

These results are echoed in similar patterns globally. A 2015 US study conducted by the computer company GFI Software found that 74 percent of those surveyed read work email on weekends, more than 39 percent are still looking at email after 11 p.m., and 54 percent checked work email during their holidays.[29] By 2017 these figures had not shifted significantly from my survey of email use in the United States, which reported that nearly 70 percent of people regularly read or write work emails outside of normal business hours, with more than a quarter saying this happens "every day." In addition, more than a third of Americans feel their workplaces expect them to be contactable outside business hours. Drilling down a little deeper I found this demand becomes higher in certain industries. Perhaps not surprisingly the occupational categories of Business/Financial; Education/Research; and Sales/Retail were most likely to feel the pressure to be always contactable by their employer.

Although many studies argue that increased figures for checking email outside work results in an overburdened and stressed workforce, the American Psychological Association (APA) maintains that such access delivers a greater sense of flexibility and productivity, which in effect "helps improve the fit between their work and non-work lives." Its 2013 investigation on work-related communication technology considered a range of work-messaging applications including email, text, and voicemail and discovered that 54 percent of respondents checked messages while on sick leave with the figure registering higher for men than women. Looking at the demographic spread of people's use of work-related communication systems outside work, it discovered that younger employees checked messages more often than did their older counterparts. Approximately 60 percent of 18- to 34-year-olds accessed out-of-work communication daily contrasted with only 43 percent of employees aged 45 to 54 who did so.[30] The difference between ages is mirrored in another study titled the "Email Overload Survey Report," which explores correlations between the rise of mobile devices and the increasing prevalence of accessing work communication outside normal hours. Where the APA study collapsed different devices and platforms, this survey focused on mobile email by asking the screener question, "Do you receive company emails on your mobile device?" It found that one in four US adults qualified for the survey, with 63 percent not accessing work emails on their phones. Income played a part; those earning more than $150,000 were the group most likely to access work emails on their mobile device.

When asked what communication platform they would choose to replace email, the majority of respondents opted for text messages. This was followed by online chat and internal intranet, with only 11.5 percent saying they wanted social media to replace email at work. Of those who chose text to replace email the majority were aged between 55 and 64 followed by those aged between 45 and 54. Respondents aged 18 to 24 were the least likely to select text messaging to replace email.[31] Overall, the survey argued, "employees squander more than a month each year checking email outside of work hours."[32]

But as another study explained, it's not simply the actual time expended out of work accessing email. The demand to respond to email may not be explicit but instead the result of work norms and cultures. These authors have identified that "anticipatory stress" is in play. Defined as a "constant state of anxiety and uncertainty because of perceived or anticipated threats" the survey argued that "organizational expectations can steal employee resources even when actual time is not required because employees cannot fully separate from work."[33] Similarly, Keri Stephens's insightful work on the implications of "reachability" is useful here. Based on more than two decades of empirical research about mobile communication within organizations, her work has examined the shifting negotiations over work/life balances. Stephens's participants revealed significant issues of power and control over reachability and, notably, unavailability that manifest with the increasing reliance on mobile devices in the workplace. According to Stephens, when "people believe that everyone around them is always reachable, it will take new strategies to help others understand that there are limits, often out of their direct control, that can make them unavailable during work hours and their personal lives."[34] More broadly, such research reflects the growing interest in "disconnection" as a significant communication strategy. Emily van der Nagel, for example, argues that the "forced connections" of algorithmic social networking systems have meant users are increasingly finding novel ways to resist such overtures.[35]

Bureaucracy and Organizational Governance

Having established the prevalence of email in the workplace, we need a critical lens through which these figures can be contextualized. As a form of communication email can't be understood in isolation but is part of the media ecology structuring an organization. When businesses express their desire to rid themselves of email, they are ostensibly yearning for a horizontal and flattened system of organizational governance. Here is where social media steps in to offer a collegial hand. Struggling with institutional hierarchy, collaborative spaces are now opened up by innovative modes of communication such as content management systems or cloud-based workflow tools. But the refrain is familiar. As mentioned, we've heard this promise before when it was made about email. This is not to say that alternative technology systems can't emerge to help organize our daily work lives or that our imaginations shouldn't stretch to think of better workplace environments for us to inhabit.

What I am suggesting is that we revisit contemporary modes of organizational design and perhaps rethink some of the assumptions that shape its interactions and encounters. Understanding how email functions within bureaucratic spaces and systems forms part of this conceptual work.

In this section I first unpack the meaning of the term "bureaucracy" and how it comes to be associated with dehumanizing workspaces. How could anyone be *for* bureaucracy? As we'll see, there are solid cases advanced for why this form of institutional governance persists. Often underpinning both sides of the debate is a dismissal of affect. Bureaucratic arrangements are devoid of emotion, instead framed by discourses of restraint, impartiality, and professionalism. In response I introduce "bureaucratic intensity" as a critical perspective by which to understand our antipathies toward email and work.

Drawing out the role of media consumption within bureaucracy, I look at some of the key sites where email frames or codifies social relations, including misdirected emails and those read in error. It's true that the standardization of policies and procedures, the compliance rhetoric of Occupational Health and Safety regimes, or the paperwork demands of an HR department resonates with us as a dead weight. Battling forms and paperwork, it is not hard to see us trapped in Max Weber's "iron cage." Yet our annoyance, frustration, and often-stoic humor when faced with the "ordering devices of everyday banality"[36] are incredibly *human* in their response. Far from merely machinic, bureaucracies help structure emotion and feeling.

Understanding Bureaucracy: Definitions and Critical Trajectories

Right from its inception in the eighteenth century, bureaucratic design as a political, economic, and social model was ridiculed. Tracing the etymology and critique of the term "bureaucracy"—a neologism comprising the French words for "writing desk" and "government"—Ben Kafka notes that although "bureaucracy" first appeared in print in 1764 in an issue of the highly respected biweekly newsletter *Correspondance littéraire*, it had been coined years earlier, in conversation with the newsletter's editor, by the economist Vincent de Gournay. Kafka points out that in addition to the three regimes—democracy, aristocracy, and monarchy—Gournay's new word defined a "fourth ... form of government, ... rule by a piece of office furniture ... to include the men who sat behind it, the offices in which they found themselves, and ultimately the entire state apparatus."[37] For Kafka, popular literature, rather than the political or philosophical treatise, is key to its dissemination. As he put it:

> In 1750, the word "bureaucracy" was nowhere; by 1850 it was everywhere. While once-common invectives such as "ministerial despotism" all but disappeared, funny or plaintive or funny-plaintive stories about "bureaucracy" flourished, spreading from France to England and Germany and, eventually, around the world. ... Literature was always one or two steps ahead of philosophy or sociology when it came to understanding the powers and failures of paperwork.[38]

The saliency of Kafka's work is how it investigates bureaucracy as a fundamentally *irrational* communication system. Although he deals with affect in rather a cavalier manner, Kafka's insight is what he calls "the psychic life of paperwork," the various unconscious frustrations and outbursts that drive us in our daily encounters with bureaucratic media. Drawing from psychoanalytic theory, Kafka shows how to "take paperwork seriously" and to highlight that those jokes and anecdotes we use to deal with the labyrinthine systems of form filling are a recognition of its power to move us inexplicably. Such a view opposes the rational, ordered, and excruciatingly explicable model of organizational bureaucracy. Faceless and automated, inefficient and grinding yet simultaneously overburdened with productivity measures, bureaucracy, as Paul du Gay also notes, has been constantly pilloried.

Writing in its defense, Du Gay shares something of Kafka's rejection of the idea that instrumental rationality constitutes the conditions of possibility within which bureaucratic structures function. Characterized as "bureau-bashing," Du Gay traces two forms of dismissal shaping the scholarship about state and institutional governance. One arises from critical theory of the sort advanced by Zygmunt Bauman and Alasdair MacIntyre, where bureaucratic social relations fragment and alienate the citizenry. People feel distanced from decision making, their actions are merely "cogs in the machine." To refute this Du Gay looks deeper into Weber to show that, far from immoral and depersonalized, Weber's bureaucracies open up new domains of ethical conduct.[39]

The second trajectory of bureau-bashing occurs in the form of neoliberal "entrepreneurial governance" that rejects bureaucracy because it stifles creativity and is blind to innovative business models. Where entrepreneurialism seeks to reform public sector organizations through market mechanisms, competition, and decentralization, Du Gay suggests their dismissal of bureaucracies misunderstands the role played by the state in guaranteeing the delivery of key services. Although this is a standard criticism of neoliberal socioeconomic frameworks, Du Gay's contribution is to reveal the messianic discursive formation that produces the charismatic manager. In this entrepreneurial persona the workplace is transformed into a site for pleasure. Emphasizing meaning, creativity, and individual autonomy, managerial communication characterizes work "not as a painful obligation imposed upon individuals, nor as an activity undertaken for instrumental purposes, but rather as a vital means to individual liberty and self-fulfillment." Managerial charisma thus unites that which bureaucracy is accused of fragmenting: it blurs boundaries between work and personal life, reason and emotion, and the public and private spheres.[40] However, along with a "god like" vision statement goes an arbitrary system of decision making because as "mystical things, visions cannot easily be rationally challenged."[41] Paradoxically then, authoritarian styles of management emerge from those very postbureaucratic discourses that had sought to replace hierarchical thinking in the first place, an observation made elsewhere in the literature.[42]

Du Gay's perspective represents one side of a debate that has been raging for at least the last 50 years.[43] Voicing the opposing view are those who see waste, inefficiency, and, in some cases, corruption enabled by bureaucracy. In his highly cited 1992 study *The Social Production of Indifference,* Michael Herzfeld argues that a "secular theodicy" drives the bureaucratic machine, a fundamentally unjust system whose obfuscating processes lead people to adopt an indifferent attitude toward suffering and inequality. Afflicting bureaucrats and the public alike Herzfeld explains that while "disgruntled clients blame bureaucrats… the latter blame excessively complicated laws, their immediate or more distant supervisors, 'the government.'"[44] Shuffled between public officers and institutions that are unable to help, bound by red tape, dispirited and disempowered, citizens lose their sense of responsibility for each other. Rather than enabling accountability, state bureaucracy fosters calculated indifference. What's crucial for me is the significance Herzfeld places on the personal, emotional, and ritualized encounters people have with bureaucracy. The systemic indifference that Herzfeld identifies occurs at the local level where forms and paperwork are used as justification for racist or xenophobic attitudes. Ralph Hummel, who is squarely in the camp that sees bureaucracy as dehumanizing adopts a similar polemic. Leaving no room for human agency, bureaucracy "rides roughshod over experience, emotion, belief, faith, purpose, meaning, feeling, judgment, deliberation, resistance."[45]

The scholarship from management studies, organizational behavior, and public administration is useful for nuancing our understanding of bureaucratic structures. For one thing, bureaucracy is not monolithic. As Janet Newman points out, within a single organization, bureaucratic elements may jostle against entrepreneurial aspects.[46] Certainly anyone working within the university sector will recognize that an adherence to formal procedures and policies often clashes with the imperative to be innovative and agile. On the one hand the university operates within a control and command, hierarchical structure, while on the other hand it demands the creation of distributed networks to achieve collaborative research outcomes. What's missing from these organizational literatures, however, is an explicit focus on the *media* of bureaucracy, the material circuits of communication and social practices that are its structure.

One notable exception is Catherine Turco's study, *The Conversational Firm: Rethinking Bureaucracy in the Age of Social Media*. In this corporate ethnography Turco spent ten months embedded at the pseudonymously named social media marketing company TechCo to map a new institutional form that she argues "transcends bureaucracy with openness."[47] TechCo, "a darling of the technology trade press," was established in about 2007 and located within an "urban region" of the United States. Characterized as one of the "fastest growing private companies" in America, it has apparently attracted an "almost cultlike following among social media marketers and enthusiasts."[48] The tantalizing ambiguity about the organization's identity is the result of a nondisclosure agreement (NDA) the author signed for the project to occur. Turco recounts her struggle negotiating a number of competing interests just prior to publication of the book. After her field work had been carried out—some 100 interviews with past and former employees; participant observation at meetings, social events, and team building workshops; and analysis of TechCo's use of internal media together with the company's outward facing social media presence—she sent the manuscript to the company's executive team. They suggested the NDA be discarded and the company identified. Part of their motivation was that TechCo itself was known to be "radically open and transparent."[49]

How could an organization recognized across social media for revealing publicly its own internal decision processes suddenly become shy? Turco grapples with fundamental questions of research ethics as she argues for the ongoing confidentiality of her participants who, she reasons, were assured the company name would remain disguised. Although perhaps they could sign a subsequent agreement to the effect that TechCo be identified, this overlooks that at the time of interview the employees had been promised anonymity. Ethically she explains she "found it problematic to modify the terms of subjects' participation ex post."[50] Still, she did need to contend with the fact that using anonymous sources often impacts negatively on book sales, and verification of facts can be called into question. On balance she opts for retaining the anonymity but speculates she wouldn't be surprised if TechCo actually outed itself on social media in the future as the subject of the book.[51]

Turco provides a useful frame through which to understand two key points, one of method and the other theory. First, as I will discuss in chapter 6, most of my Enron research participants opted for complete anonymity or pseudonymity as the basis for their responses about the publication of their emails. Persuading people to speak about their experiences at Enron was conditional on their anonymity. A number of people were unequivocal in their desire to remain anonymous; expressing they had no wish to "drag up the past" of Enron so that any casual internet search could bring it rushing back, both symbolically and materially, into the present. As one participant put it while deciding on a pseudonym: "Google searches are now a way of life and I'd prefer some of the more relevant things that I have been doing recently to be at the top of the search list." In other words, participating in my research could make their involvement with Enron visible again.

The point of my study is to show the effects of that visibility, not to generate it. Yet even as I created pseudonyms to describe their former company positions and current roles in the energy and legal sectors, I worried about potential identification. If I strayed too far from their original connection with Enron, it was difficult to argue for the validity and integrity of the findings. But just as importantly, as Turco notes, participants should be able to recognize themselves in the subsequent representation, their experience of reading the research publication ought not to be an "alienating" one.[52]

More broadly, I'm interested in what Turco's work can tell us about the theory of bureaucracy and its everyday instantiations in organizational life. For Turco, and her TechCo participants, bureaucracy is diametrically opposed to openness and transparency. A central claim of her book is that TechCo embodies a new institutional structure, what she calls "the conversational firm," which is able to confront "the tradeoffs of openness and bureaucracy directly" resulting in a "radically more open communication environment than we have ever seen before."[53] What the desire for radical openness might actually mean is particularly salient in the ethnographic work she presents about TechCo's "visceral hatred of HR." Astoundingly, to my mind at least, TechCo had eschewed an HR department when it began and then, even when its workforce grew to more than 300 people, had managed to resist formal employment protocols for many years to follow. According to TechCo's founders, "human resources was a rotten legacy of 'old school' corporate bureaucracy, a source of unnecessary rules and processes that directly threatened the sort of openness and flexibility" at which they were aiming. Instead, the chief executives "simply refused to participate in the ceremony of it all" despite repeated calls from its board of directors and later its own staff.[54] With no formal maternity leave provisions, no proscribed procedures for termination or hiring, and unclear performance indicators for promotion it is unsurprising that the employees of TechCo called for reform. What *is* surprising is that anyone would expect the management of a Silicon Valley start-up to behave differently. Why wouldn't they want to do whatever they liked, unencumbered by the irritating requirements of workplace employment rights? Grudgingly accepting the staff wishes, the executives of TechCo finally installed an HR officer only to appoint above her a manager with the "directive to block HR practices that might threaten the company's existing culture."[55]

At the same time that Turco was working hard to protect the anonymity of the company she had called TechCo, another author was busy publishing his "tell-all" book on the very same organization. It turns out that cloud-based marketing company HubSpot was the basis of the two books, both published in the same year. HubSpot is based in Cambridge, Massachusetts, with worldwide offices in Bogotá, Paris, Dublin, Berlin, Singapore, and Sydney. Employing over 2,500 staff, in 2014 it floated the company on the New York Stock Exchange with its total 2017 revenue recorded at $375.61 million. In 2018 it launched "HubSpot Ventures," a $30 million fund to support marketing and analytics start-ups.[56] As noted in the previous chapter, HubSpot is one of the leading US email providers, and its services include search engine optimization, data analytics, and email marketing automation. After a number of years working at HubSpot, the well-known tech writer Dan Lyons wrote his exposé, its path to publication apparently mired by strenuous efforts from the company, including extortion and email hacking, to prevent its release.[57] The book, titled *Disrupted: My Misadventure in the Start-Up Bubble* describes the company culture as "frat house meets cult compound," and the experience also inspired Lyons's script writing for the hit HBO program *Silicon Valley*. It is certainly not hard to recognize the same company in the pages of these two books, as both authors, to varying degrees, attempt to grapple with the efficacy of bureaucratic design and "radical transparency" playing out in the software industry.

Bureaucratic Power and Gender

While issues of power have been echoing quietly through my analysis of bureaucracy, it's now time to look more closely at how structural or systematic governance shapes certain modes of action and voice. And it is feminist analyses that help reveal productive lines of resistance and point to how we might understand the function of bureaucratic intensity. As would be expected, there are different articulations and historical approaches in feminist thought regarding bureaucracy.

Although there is not adequate room to do justice to the wealth of research conducted on how gender registers in workplaces and organizational structures, I want to draw out from the literature a set of questions that can help make sense of how bureaucracies shape social relations. Within many of the sources traced above, feminist perspectives remain largely ignored, their appearance relegated literally to a footnote. In Du Gay, for example, feminist studies are collapsed under the heading of "bureau-bashing" along with entrepreneurialism, and dismissed because they are driven by a quixotic desire for unity. Across the critiques of bureaucracy, he argues, "Certain shared assumptions are discernible. Chief amongst these is a thoroughly romantic belief that the principle of a full and free exercise of personal capacities is akin to a moral absolute for human conduct."[58] Driven by this belief, opponents of bureaucracy demand "all areas of life should be united, and that the individual overcome the alienating distinction between the different social roles she is forced to inhabit."[59] Rendered invisible (and made all the more grating by the disingenuous use of the feminine pronoun) is women's unarguable and persistent inequitable treatment in the pay structures of bureaucracy. This oversight is striking considering that Du Gay himself is so articulate on how bureaucracy as a socioeconomic system can ensure fairness. If "having it all" can be seen as a version of Du

Gay's romantic unity, then perhaps he is right; postfeminist critiques and "entrepreneurial feminism" have certainly identified that neoliberal markets impact negatively on women's lives in the workforce and at home.[60] Obviously, I'm being facetious. That is not his point. Nevertheless Du Gay's rhetorical move, that sly "politically correct" gesture, is illustrative of how women are often made to do the work (narratively, materially, emotionally) of others while they themselves remain unable to reap its benefits. It is the last register I want now to explore—how the emotional sphere gets overlooked in bureaucratic communication and how to bring the intensities of email back into the equation.

We have already uncovered some of the ways affect gets ignored in the social theory of bureaucracy. Usually this is achieved by pointing out it's a straw man argument that falsely opposes the rational, ordered, organizational system to its other, that of feeling, intimacy, and emotion. Such studies, however, rarely go on to discuss the actual emotional labor in bureaucratic spaces. Another critical stance rejects bureaucracy *because* it is unemotional and dehumanizing and so necessarily cannot see within its frame how to understand organizational care and intimacy. Feminism has a long tradition of evaluating affective labor—recognizing and bringing to the surface the myriad ways that caring, kindness, and support shores up the formal structures of economic and social relations because it adds surplus value.[61] Not only do women often provide primary support in the home, but when care is required in the workplace this is also devalued. And where care itself is the service provided, with allied health and aged care for example, women and people of color disproportionally dominate the underpaid and under resourced sectors.[62]

For many feminists, bureaucracy is fundamentally at odds with the lives of working women because it controls emotion and undermines principles of caregiving. Reflecting on the potential for public administration reform, Camillia Stivers writes:

> Theorists may extol the virtues of the responsive, caring bureaucrat who serves the public interest but the argument will face uphill sledding until we recognise that responsiveness, caring and service are culturally feminine and that, in public administration, we are ambivalent about them for that very reason.[63]

Similarly, for Kathy Ferguson, the subordinate actors in a bureaucratic system adopt feminized traits including sensing both the mood and prejudices of managers, learning to please, and placating adversaries or smoothing conflict situations so that others are seen in the best light and helped to excel.[64]

Although her ground breaking 1984 work *The Feminist Case Against Bureaucracy* faced critique for too starkly stating the problems inherent in bureaucratic governance and for offering little scope for change, it decisively carved out a critical space to discuss gender and workplace structures.[65]

"Reply All" Disasters and Bureaucratic Intensity

One response to Ferguson has been to canvass alternative organizational environments, predominantly nonhierarchical and collaborative decision-making systems.[66] My contribution—somewhat more modest but no less salient, I hope—is to insist on the affective aspect of workplace email. I am interested in how the rational and ordered register of bureaucracy is shown as leaky when through its systems of process emotion trickles and seeps. Bureaucratic intensity describes a register of office communication that acknowledges modes of rational, professional behavior while recognizing how moments of fury, gestures of compassion, words of support, practices of listening, and sighs of frustration also frame our workplace encounters. By bureaucratic intensity I mean to capture both the everyday banality of the office and the dramatic significance of affect.

In bringing these two logics together I draw on the insights of Anthony McCosker, whose work helps to frame the myriad affective flows, both grand and minute, that drive us to think and act. As he puts it, "The word 'intensity' seems to have its own gravity or inertia, but nonetheless incorporates more subtle 'microscopic' movements of force."[67] Such currents ripple across many of the narrative survey responses as the participants recount their experiences of workplace email. Bubbling to the surface and barely kept under check are feelings of irritation, embarrassment, intimacy, surprise, humor, pleasure, exhaustion, and regret as people reveal the details of their email communications at work. In some situations, tempers flare as the result of mistakes made with the technical conventions of the software program. Yet these encounters are often experienced beyond the merely instrumental. The inadvertent use of the "cc" field, for example, can have far-reaching consequences, as one of the participants discovered. When asked to describe a regretful email encounter, one respondent wrote extensively about the experience:

> An employee had complained about an issue and I didn't notice that another manager had included her in his email to me, and other senior staff, about it.

> I expressed the opinion that if the employee needed her hand held to the extent the email indicated, she probably wasn't right for the organization. She subsequently resigned. I don't regret the opinion—which I believe was right—but I do regret inadvertently expressing it to the employee in such an unvarnished way.

Bristling with bureaucratic intensity, this account must surely resonate for any of us who have mistakenly copied a coworker or witnessed such an event in an email chain. Beyond such recognition, however, is the experience of the former staff member who has been doubly insulted. First is the issue that drove her initial complaint in the first place, and then comes the unexpected and skin-crawling realization that she is the ridiculed subject of an ostensibly confidential discussion.

The emotional fallout from misdirected workplace emails was a recurring theme in the survey responses. One respondent wrote of the regret at having sent "a negative email about an individual whose manager was accidentally copied into the distribution"; another confessed to hitting "reply all" instead of "forward," an inadvertent action that "burst open" the discussion about the recipient being "the root cause of the problem." Another respondent described feeling highly embarrassed at having hit "reply all" when the "content was meant for one person only." Although others probably regarded the actual email content as innocuous, to this person it felt quite shaming to have so publicly failed to master the protocols.

Bureaucratic intensity is not always the work of a moment, the result of inattention to the software conventions of an email program. Instead it may erupt after periods of workplace resentment have been building, and its refrain will then linger. Asked to describe email use in the workplace, one respondent explained:

> I became very frustrated and angry at work over lack of responsibility taken by one of the suppliers I deal with and I sent an email to their business development manager that was abrupt. Had I been calmer I could have been less short yet still conveyed my irritation and disappointment without seeming quite so scathing. The business development manager no longer corresponds with me unless he absolutely has to which has made dealing with this company much more difficult.

For some people, the affect of workplace email was not necessarily dramatic. As many of us can attest and discussed above, even fleeting moments of feeling can significantly shape the experiences of our work environments. Explained one survey participant, "I hadn't finished editing something and sent it by mistake. Not a terrible error but embarrassing";

while another commented on the regret they experienced when they "sent work email to a wrong client."

Although most of the responses dealt with feelings of annoyance, regret, and anger expressed through email communication in the workplace, there were instances of levity as people discussed how humor and banter helped to produce intimacy during their email exchanges. In one instance, the shared understanding of the perils of email saved the day.

> Technically...the email wasn't a mistake, the mistake was that I sent it to the wrong person. It was supposed to go to a friend but instead it went to an Instructor, thankfully he was a really cool and crazy (fun) guy so he laughed at it and told me don't worry about it. The message was for a friend so we call each other names and all of that so it was pretty bad.

Similarly, another person admitted: "I accidentally sent a few pics of plants looking like certain body parts to everyone in the Western District of this company I was working for.... I felt so glad the District Manager had a great sense of humor." But simply designating a message as funny does not necessarily exclude the possibility of rancor or at the very least ambiguity, particularly when used across hierarchical lines. As one respondent confessed about a mistakenly sent message: "It was a joke email but my employer definitely didn't find it amusing."

The Role of Emoji in Workplace Vernacular

Thinking about how humor and emotion might play out across the circuits of communication that structure our workplaces raises the affective role of the emoji. Whether it is winking, smiling, holding thumbs up, applauding, wincing and crying, or expressing disbelief through the face-palm, the emoji has become an important aspect of workplace vernacular. In chapter 3, I discussed efforts by email service providers and marketing companies to monetize and encourage its uptake. Now I want briefly to explore people's attitudes to its use in business contexts.

Respondents to the US survey were largely critical of its role in the workplace. Chief among its disadvantages was that use of emoji signaled "unprofessional" behavior with terms like "juvenile," "childish," "adolescent," and "tacky" peppering their descriptions. Explained one participant, "Emojis are meant for fun and it's not professional for work"; "it's work not play," added another. While most felt strongly that at work "you should be keeping it professional and emojis are not on a professional level," some respondents reasoned that context played a part, particularly that of the relations existing between coworkers: "it depends on the relationship you have with the person you're sending to and how formal the situation is," with another person noting that emoji are "OK with casual emails between close colleagues, but should be used sparingly." A particularly thoughtful reply came from a participant who suggested that the volume of emoji might also be significant because some people are known for overuse: "Indicating tone is an important part of communication and language. Responsible use of emoji should not be a faux pas. For example one per email." Tone came through as a major driver for the choice of whether to use emoji. But interestingly people often gave the same reason for either rejecting or embracing the use of emoji within email. Tone was either enhanced or hindered by its inclusion. Questions about whether emoji was appropriate for work email produced a high volume of descriptive responses:

> Emojis are a good way to convey tone.
>
> I think they convey emotions better than just words do.
>
> An emoji helps explain your feelings.
>
> In general, I don't think it's appropriate—not professional, but there may be rare occasions they would be helpful to get the point across.
>
> It sends humor to someone who might be having a bad day.
>
> It makes email more personal.
>
> No, it adds emotions.
>
> I think it depends on the seriousness of the business. Some coworkers you might be more personal with, others should remain strictly business. Real professionals find a way to make the smile in their voice be heard through the text without emojis.
>
> No, because emojis indicate feelings.
>
> If it is a serious job, you should avoid putting unnecessary excess ... in your messages and get straight to the point.

Again, context is crucial, and the hierarchy that exists within workplaces plays a significant role in people's emoji use, as well as the type of organization or occupation. One person reasoned that it "depends on the type of business. A dentist office could use a happy tooth, a bakery could use a cake or cupcakes, but a government office should not use any." Someone else explained that emoji use will differ according to whether you are sending it to "a superior, equal, or subordinate," and another noted they have never used emoji because they work within the "corrections department and the Prosecutors Office." Age has a part to play with those in the 35- to 44-year-old bracket appearing the most certain that emoji was not suitable for work email contexts. Conversely, those in IT and sales figured high among those who felt emoji use in work contexts was appropriate. The last word on whether emoji is suitable for workplace email has to go to the respondent who replied, "No, we do not have a sense of humor in the Federal Government."

Conclusions

In this chapter I have sought to uncover what it means that email still dominates our workplaces. Across numerous surveys people report that email is chosen as the most frequently used method of workplace communication. Praise for the newer collaborative suite of apps like Slack or Twist abounds because of the nonhierarchical structure these programs offer by encouraging team work in ways that email is not apparently equipped to do. But as I discussed, the very same claims were made about email as it began to enter the workplace decades ago. Writing in 1992, one commentator stated "corporate e-mail is breaking down hierarchies by making upper managers more accessible."[68] This suggests that workplace organization and governance is often central to how new workplace technologies are understood. Discussing the major issues raised by workplace email I identified five broad categories, each representing a constellation of associated practices, technologies, and cultures: governance (covering organizational structure, hierarchy, and labor relations); productivity (referring to information overload, interruptions, work/life balance); monitoring (which includes encryption, information security, privacy, legal liability, or exposure); device (covering distributed workplaces, mobile email, cloud-based email, media choice); and affect (referring to negative and positive emotion, employee conduct, and emoji use). If these are the broad domains of interest structuring the literature on workplace email, I then looked more closely at demographics of usage.

As I mentioned in the previous chapter, quite often interpreting these results means tracking down the sources of funding for a particular study. Recognizing the situated nature of a study does not necessarily invalidate its findings. On the contrary, stitching together the threads of economic or material investment actually enriches the data to bring important context to the analysis. Nonetheless it has been exceedingly difficult to find independent figures on email use, a deficiency I attempted to address with the two international surveys I conducted. These indicated that email is the preferred mode of digital communication at work. Beyond the figures, I was interested to learn more about my respondents' perceptions, beliefs, and habits across the range of email practices evident at work as they admitted to their embarrassment of misdirected emails, or the irritation of excessive emoji use.

As I began to analyze the survey responses using the scholarly material on workplace affect, I wasn't quite satisfied with the critical offerings. Instead, I wanted to find a theory or system of workplace organization that could match email in its everyday or routine nature. Or put another way, what term might describe all the categories of workplace email practice enumerated above, in particular governance, productivity, and affect? Bureaucracy, with its almost invisible or taken for granted processes, eloquently captures what's often at stake in our institutional forms of life and labor. So, I canvassed the critical terrain of bureaucratic perspectives by pointing to its central debates. Here I noticed something of a gap because the strength or value of care and emotion at work is often overlooked. Now, perhaps bureaucratic organization may seem, to many readers, to be an oddly quaint topic for critical application, the province of dusty pen pushers or file-obsessed archivists. But it has recently experienced something of a renaissance as many try to think through "radical decentralization" while grappling with the Silicon Valley start-up or "disrupter" and its associated forms of institutional design.[69] And, as I showed, the email service provider HubSpot has become (willing or otherwise) a test case for the efficacy of "postbureaucratic" systems of governance.

While the slightly sordid travails of a tech start-up are no doubt fascinating, the main point of their inclusion was to show how bureaucracy reoccurs through much of the research on contemporary organizational governance. In turn, this was the frame through which I interpreted the affective dimension of workplace email found in my survey respondents. In particular I suggested that the phrase "bureaucratic intensity" might help to capture both the everyday banality of office life and the dramatic dimensions of affect. But as I discussed, while some respondents revealed their open hostility to or frustration about workplace email and would detail coworkers' digital illiteracy for example or bosses' failure to respond to requests in a timely fashion, affect was also expressed in playful or whimsical ways. Emoji use was a site for energetic debate, and although most respondents did not agree with its use in professional settings some admitted to finding it useful in the workplace. "Email makes it difficult to provide emotion, so can be misconstrued. Emojis can help provide context," wrote one respondent.

We've seen how the "reply all" disaster can cause regret and shame in its participants because of the suddenly public nature of its contents—with "suddenly public" anticipating the discussion of the Enron and Hillary Clinton emails in chapter 6. Next however, in chapter 5, I look at email lists and groups as one of the most enduring forms of email communication. Again, affect and labor are in play as we trace the role of group moderation and governance.

5 Moderation and Governance in Email Discussion Forums

Writing in 2017 about his time spent moderating the email discussion forum Cybermind, Alan Sondheim admitted, "I was a boor." With characteristic sensitivity and the intuition of an experienced digital ethnographer, Jon Marshall is quick to point out that any blame for such possible shortcomings lies more with the platform, its software constraints, the relative novelty of its communicative format, and the collective norms of the group than it does with the skill set of a single person.[1] Marshall is the author of *Living on Cybermind: Categories, Communication, and Control*, a comprehensive and finely grained investigation of this mailing list spanning the period from 1994 to 2007, in which he looks at the forum's modes of governance and the daily interactions of its subscribers. Through the conceptual lens of anthropology Marshall shows how the list is shaped by the flows of the gift economy, a desire for authenticity, and the interplay of "asence," a key term he coins to analyze the fluctuating perceptions of online presence and absence with which Cybermind participants often grapple.[2]

Reading Sondheim's reflections about his work as a moderator I was somewhat perplexed. I too have conducted research about Cybermind, published in my own book on the relation between nineteenth-century letter writing networks and the mailing list cultures of their twenty-first-century counterparts.[3] Through his posts to the group, Sondheim had always appeared sure-footed in his moderation duties: level-headed about what kinds of norms should be encouraged for the group's cohesion, and in turn, what conventions were not helpful in generating productive online interaction. But to enact that authority obviously comes at a price. For Sondheim there's a disconnect, perhaps, between what the role demands socially and technically and his ability to fulfill those requirements. The candor of his confession also piqued my interest because it helps to highlight the affective dimension of online moderation.

Although social media governance has gripped the critical imagination of communication scholars, less has been said about its provenance through online discussion forums. Yet many of the recent studies conducted about the digital labor of curation, site administration, and content moderation deal with issues that have already arisen during earlier periods of internet governance and development. In this chapter, however, I am not primarily concerned with the lens of history; understanding the historical paths of email was the work of part I, where I focused in detail on the pivotal role that early design decisions and the 1990s communities of practice played in shaping our patterns of email consumption. Now I want argue for the contemporary significance of email discussion groups and forums. In so doing I will bring to light some of the key issues facing these conversation spaces, exploring the labor of moderation and how this is inflected by affective flows.

I begin by explaining the social and technical definitions of different types of email lists. In turn these various software capabilities open up questions of archival policy and the commercial uses of email lists through the issue of spam. Contemporary email discussion lists are a vibrant resource for support, collegiality, information exchange, and advocacy across diverse populations, so I also discuss their impact broadly in the literature before moving on to examine the specific role of moderation. Here I draw on email interviews I conducted with 20 moderators of discussion lists spanning a diverse range of groups and topics including professional development, mental health support, academic research communities, sports, and hobbies. Responses are placed within the context of three categories: affective management, perceptions of labor, and the future of email discussion lists.

Different Types of Email Lists

Before exploring how moderation and governance function, let's look briefly at the diversity of social and technical practices that are captured by the term "email list." Descriptions like mailing list, newsletter, discussion forum, listserv, and email group tend to be used interchangeably, blunting the precision with which a researcher might conduct analysis. From the perspective of user experience, these applications can appear to operate in similar ways, yet the underlying software differs. The most obvious example is an email list manually compiled by collecting email addresses and entering them into the "cc" field versus an email list hosted by automated distribution software accessed by subscription. In both cases users simply hit the "reply all" key to email the entire group. But with automated management systems the mail is received by a single address, sometimes called a "reflector," or more commonly "list address," that organizes the distribution function. Popular list automation software includes Listserv, GNU, and Mailman. Because Listserv was the earliest form of proprietary email list management software, the term is often used generically to refer to discussion groups in general. Email groups or lists may also be compiled manually by entering recipient addresses into the "bcc" field, meaning addresses are not visible to all those subscribed to the email group (as is the case with automated distribution software) so that "reply all" only sends an email to the administrator of that group, which is unlike the case with list software.

If these gradations seem insignificant or overly fussy in their level of detail, it's because such material features of a communication system regularly have an impact upon platform governance. A list moderator often plays an intermediary role between the software constraints of a given system and the voice of a particular community. Moreover, both manually organized email groups and those relying on distribution software provide important tools for advocacy and grassroots networks. In her research about the participation spaces of citizen-led community groups and activist networks, Ella Taylor-Smith found a wide range of email lists were deployed effectively.[4]

Public Access and Archiving

Public access varies considerably across different types of email groups, with some applications automatically archiving list posts and providing searching and download functionality while other email lists restrict viewing privileges to those included in the distribution list. But just because an email group uses automated distribution software doesn't mean the posts will be publicly archived. An instance of this occurs with the internal communication systems of an organization that uses a single point of email management where the email address carries the syntax of "mediastaff@lists.swin.edu.au" (let's say) or "allsales@maillist.fashionforward.com." In these cases, the email addresses function as lists requiring approval by administrators before emails are disseminated to relevant staff in the organization. Although emails could be archived internally by the list administrator, they are not available for public searching. This differs from the situation where email lists connect people across institutions and sectors rather than from within a single organization.

Publicly accessible email discussion lists can be formally and informally constituted; their user base spans community groups, research organizations, standards bodies, software developers, health practitioners, cultural heritage sector members, and academic groups. The list archive of the Association of Internet Researchers (AoIR), for example, dating back to May 2001, is searchable via email topic, author, and date of posting.[5] For Axel Bruns, Immediate Past President of AoIR, the publicly accessible feature of the archive is crucial:

> Air-l is a public discussion forum, and we believe that the exchanges here are of broader use to Internet researchers and other interested parties—in particular perhaps when our members are collectively compiling reading and resource lists on particular topics.

The mailing list archives for Wikimedia, the not-for-profit organization responsible for Wikipedia, are publicly available but the search capacity is quite limited for nonsubscribers.[6] Finally, the work of the Internet Engineering Task Force (IETF) is carried out substantially through its mailing lists. There are hundreds of lists hosted by the IETF encompassing standards literature and policy development, with its archives having recently upgraded to include sophisticated search capacities.[7]

Archives of email discussion groups are a vital aspect of public domain knowledge. From providing historical records of momentous events—such as the terrorist attacks of September 11, 2001—to sharing research resources and connecting interest groups across the globe, mailing list archives play a central role in creating and sustaining public-sphere discourse. For enduring lists like VICTORIA, the forum to discuss nineteenth-century Britain (covering the period from 1780 to 1918 and established in 1993), the archives are an extensive resource for teaching and research.[8] As its founder, Patrick Leary, explains:

> These archives are very important. ... Not only do they contain a wealth of practical scholarly information from a wide range of experts, but the quarter-century of often very detailed discussions included in them constitute a kind of history of their respective fields. On VICTORIA, there have been a great many exchanges about teaching methods, for example—what has worked in the classroom and what hasn't, how best to teach this or that text, experiments with reading Victorian

> novels in their original serial formats, or with using digitized periodicals, and much more. Those discussions have taken many forms, from exchanges beginning "Does anybody know of a good poetry anthology?" to "How have other people taught Dr. Jekyll and Mr. Hyde?" to the inevitable "Can someone recommend a nice short Victorian novel for my class?" ... Authors, editors, and annotators of textbooks and anthologies have joined the fray, as well, finding no shortage of advice and opinion from the teachers their work seeks to reach.

Yet email list archives can be notoriously unstable. List moderators and owners frequently need to address insecure hosting arrangements, funding constraints, or the outbreak of hardware problems that suddenly render an archive temporarily inaccessible or, worse, lost altogether. In response to the perennial challenge of securing reliable data storage and retrieval mechanisms, large-scale list archival services have emerged including Gmane, MARC (short for "Mailing list ARChive") and the Mail Archive. Although, as mentioned, individual list software usually provides archival capacity, these aggregating services offer a more stable hosting environment. In addition, the searching features enable interested people to examine huge databases by topic, author, or date. MARC archives approximately 3,500 mailing lists while the Mail Archive houses over 1.3 million postings across 4,329 active mailing lists.[9] Given such coverage this latter archival service has often found itself the target of take-down notices from the DMCA (Digital Millennium Copyright Act), a situation that has made it necessary for Mail Archive to implement various policy approaches outlined on its website.

Commercial Mailing Lists, Newsletters, and Spam

The term "mailing list" also refers to commercial marketing activities where companies obtain or compile lists of potential customers to which they send advertising material, newsletters or "email blasts." Here an important distinction is often made between third-party lists of addresses and those generated by a business from within their own customer base. As we learned in chapter 3, email marketing constitutes one of the highest volumes of email traffic, and so the sector must continually fend off accusations of spam-like behavior. Indeed, it is hard to ignore industry figures in which the categories of advertising and spam regularly top the lists that rank "most frequently delivered types of email."[10]

Spam, or junk mail, is the practice of distributing unsolicited mass email often containing harmful malware, attachments, or other embedded links.

There is considerable disagreement about how spam should be defined. Confusion arises about whether it "constitutes any unsolicited electronic mail," "unsolicited bulk e-mail," or "only unsolicited commercial e-mail," with some commentators and policy makers insisting all three conditions must be met for it to count as spam.[11] But the international nonprofit organization for spam research, known as The Spamhaus Project, is unequivocal about the definition:

> Spam is an issue about consent, not content. Whether the Unsolicited Bulk Email (UBE) message is an advert, a scam, porn, a begging letter or an offer of a free lunch, the content is irrelevant—if the message was sent unsolicited and in bulk then the message is spam.[12]

Figures differ on the extent and impact of spam. Industry data suggest that global spam volume accounts for between 70 and 80 percent of total email traffic. But many reports show there is a downward trend emerging as these rates steadily decrease year by year.[13] Spammers work by harvesting email addresses using software that scans for the @ sign across a variety of online sources including social networking sites, mailing list archives, forums, online comment fields, and chat programs. Spammers may operate by purchasing email list addresses from companies specially set up for this reason. Here again we encounter the blurring boundaries of what constitutes spam: the practice might contravene acceptable use policies and legislation, but it could actually function in a legitimate manner because the person receiving the email agrees to receipt. For some email service providers there's no such thing as a legitimate email address list. As Mailchimp warns in its "Common Rookie Mistakes" guide about purchasing lists:

> By now, everyone should know better than to buy a "totally legitimate list of 30 million opt-in emails" from a sketchy piece of spam they found in their inbox. That's pretty obvious, but there are still some vendors out there selling "opt-in" lists the old-fashioned way. They collect email addresses and ask members if they'd like to "receive special offers from third parties." Then, they sell those email addresses to other senders. It's not technically illegal, but many ESPs—Mailchimp included—prohibit sending to purchased lists. Mailchimp is a strict permission-based newsletter delivery service. This means we do not provide, sell, share, or rent lists to users, nor do we allow purchased, publicly available, third party, or rented lists in our system. No exceptions![14]

This jaunty advice belies decades of acrimonious, always contentious, and (in the end) intractable debate occurring between the commercial email sector and anti-spam lobby, a terrain expertly charted by Finn Brunton.

Perhaps not surprisingly, the lineage of spam ties it directly to the use of email distribution lists. As mentioned in chapter 2, Brunton identifies that the "first protospam" message dates back to the early days of email when marketer Gary Thuerk and engineer Carl Gartley sent out an advertisement through the ARPANET in 1978 for the computer company Digital Equipment Corporation (DEC). In Brunton's words:

> DEC had a strong business presence and established customers on the East Coast of the United States but had much less presence on the Pacific side, and it was Thuerk's idea to take the printed directory of all ARPANET addresses, pick out the ones belonging to West Coast users, and let them know that DEC was having an open house.[15]

According to Brunton this event set the scene for assessing the scale, scope, and true nature of spam over the decades to come; conceptual limits that today still confound us. One of the problems is that spam describes such a diverse range of practices, technologies, and beliefs: a "single word ... with remarkable properties" that can be "applied contemporaneously to projects with completely distinct technical means and social motives" and then "retroactively used to dub behavior on antediluvian systems whose properties would be virtually unrecognizable to someone casually retweeting an Instagram image."[16] Brunton's key insight for my purposes is to show how the community ethos of those early forums and discussion lists created a space where spam could flourish, where trying to curb it through anti-spam legislation somehow became akin to preventing free speech. "The government however good its intentions, should not strangle electronic commerce at birth" was how the Direct Marketing Association (DMA) reacted to policy interventions in 1998 aimed at reducing spam. This response, Brunton explains, was "typical of the DMA's statements, as they tried to develop their own systems for regulating spam without making it more difficult for their member organizations to send ads."[17]

Since the 1990s, spam has attracted multiple legal and regulatory responses globally. Most jurisdictions have enacted anti-spam legislation making it an offense for organizations to send unsolicited communications without the consent of subscribers. As Mailchimp notes, companies will often stipulate an opt-out mechanism in their mailouts whereby recipients are able to unsubscribe from within the body of the email.

The first US legislative solution to spam was introduced under the Controlling the Assault of Non-Solicited Pornography and Marketing (CAN-SPAM) Act of 2003. Right from its inception it was beset with difficulties, chief among them the detractors who saw CAN-SPAM as a toothless tiger, a box-ticking exercise that appeared to respond to consumer protection needs but, in reality, did not constrain companies to change their business models. Much was made of the message implied by its acronym: many joked that it actually meant "you can spam," an interpretation supported in large part because the act did not provide for the "opt-in" function, a feature of stronger laws operating in other jurisdictions at US state levels and internationally. In the United Kingdom, Australia, Germany, and Italy, for example, anti-spam regulations stipulate that email marketers may only send material to people who have explicitly agreed to receive it and, further, these companies must provide evidence for that consent. In these jurisdictions, in sharp contrast to the US CAN-SPAM Act, merely including an opt-out or unsubscribe link is not enough. As compelling evidence of consumer dissatisfaction, in my US survey, nearly 70 percent of respondents say they have actively reported spam emails.

As the CAN-SPAM Act rolled out, there was widespread speculation across the communications industry that the marketing sector had lobbied government because they feared the detrimental impact the (often quite strong) US state level laws would have on their revenue streams. What they apparently wanted, and were granted, was a mechanism that could preempt or supersede state law, in effect meaning businesses would only have to provide the unsubscribe link in their email marketing campaigns.[18] "Suddenly," as Brunton says, "it becomes the responsibility of everyone with an email address to opt out, over and over; direct email advertising is presumed legitimate until proven otherwise." Moreover, the "unsubscribe" button could often function as a trigger for even higher volumes of spam. Because clicking on an unsubscribe link alerts illegitimate marketers that the address is live, "less scrupulous marketers could take advantage of this legislation, leaving the recipient always unsure as to whether they were removing themselves from one list or putting themselves on many, many more."[19] The CAN-SPAM Act has largely been deemed ineffective. In a major impact assessment review of the act published in 2016, the authors found little evidence of its success at reducing spam rates mainly because its provisions are generally unenforced. Specifically, the study argued that imposing fines mattered little to large-scale spamming outfits that could simply factor such costs into their budgets. But lengthy prison sentences, they maintained, could prove effective.[20]

Meanwhile, in 2010, Canada introduced one of the world's most aggressive anti-spam laws.[21] Both in the penalties it can impose and the high thresholds that marketers must meet, Canada's anti-spam legislation (CASL) has been greeted with both praise and condemnation for its rigorous approach; a reaction in large part determined by where you sit in the market. Consumer protection advocates applaud the limitations it has placed on the invasive and unethical tactics of digital marketers while many businesses, particularly those located in the United States, scramble to attain compliance and worry it paves the way for more restrictive spam reforms to be enacted in the future.

But let's be careful not to draw too sharp a division between these groups: commercial interests and advocacy can function in dialogue. And anti-spam legislation and other spam preventative tools have adversely affected the not-for-profit community sector. Though some of the legislative approaches offer exemptions for registered charities, many nonprofits or NGOs, advocacy groups, and charities need to be anti-spam compliant. Not only does this mean making sure they meet their statutory obligations, it is also about being aware of clickthrough rates and deliverability scores so that newsletters and campaign messages are not treated like spam and thus fail to reach their target inbox.

Email lists still figure highly within the not-for-profit sector. A 2018 study conducted by the Nonprofit Technology Network found that 28 percent of all online fundraising comes directly from email campaigning. The research analyzed donation and engagement patterns of 154 nonprofit organizations across email, web traffic, SMS, Facebook, Twitter, and Instagram. It found that per 1,000 email addresses, nonprofits had an average of 84 mobile number subscribers, 474 Facebook fans, 186 Twitter followers, and 41 Instagram followers.[22] Despite promising figures, industry research has shown that the average nonprofit organization will lose over $24,000 in revenue a year because of spam. While size of lists and clickthrough rates affect the capacity to raise money, so does "deliverability": when spam gets stuck in spam folders it forms a significant barrier for charity organizations.[23]

You might be thinking that this analysis of spamming and the diverse markets it opens up belongs earlier in the book with the material about the email industry. Security software manufacturers and data analytic companies, together with the regulatory bodies that shape the passage of spam, all find a place within the email service provider field that I mapped in chapter 3. Here, however, I am concerned with showing the different ways that a particular email affordance—that of sending messages to groups of recipients—is understood. When we see the term "email list" it can refer to commercial, personal, and social practices—and it's important to outline these different types. Collapsing these media distinctions is one of the ways email usage gets overlooked. So far, email lists in this section have referred to a practice where, in general, the subscribers did not talk back to the list. Now let's return to the category of the email discussion list. Since participation is often a key element, how these lists are organized and governed comes into focus.

The Impact of Contemporary Email Discussion Lists and Groups

Email discussion lists continue to play a significant role in knowledge transfer, resource sharing, and professional development for research communities across diverse fields of academic endeavor. In a large-scale longitudinal study of Spanish arts-, science-, and education-based email discussion lists, Cristina Faba-Pérez and Ana-María Cordero-González found evidence for the sustained vitality of this communication channel. Between 2002 and 2011 the authors applied bibliometric modeling to their corpus of 107 different email discussion lists to gauge impact and productive value. Specifically, they explored whether it was possible to identify core academic lists using similar techniques to those that index the most influential and highly cited journals. The authors concluded that discussion lists operate as valuable indicators of major fields of interest and, moreover, offer guidance for research providers. As they explained, it is "recommended that librarians know the core distribution lists in each subject category... when forming the library collection." According to their study "electronic mailing lists of the documentation, education and medicine categories underwent growth over the 10-year period... and have a solid future prediction in this line."[24]

Similar work exists on the quality of information on academic lists and the trust invested in these research resources. In a series of studies, Uwe Matzat investigated how academic mailing list subscribers evaluate the worth of messages circulating through their email networks against criteria that includes volume of posts distributed; the perceived correlation between the message and the purpose of the list, or what gets called the "off topic" problem; and how well posts contribute to the progress of knowledge within a given field. Matzat found a key factor was the degree to which the participants' interactions were located within existing offline research networks that form through regular conferences or other collaborative projects. As he explained, those email lists, "embedded in a well-integrated academic community, provide information that is regarded as being of a higher quality" than those whose participants rarely meet. Reputations are established within these embedded networks so that the information circulated through the lists are valued and trusted.[25] Moreover, in disciplines like the social sciences where research findings are dispersed across numerous journals, mailing lists can perform important filtering, recommendation, and orientating functions.[26]

Mailing lists represent a significant platform for documentation and development within the open source software (OSS) community.[27] Across lists like the Linux kernel mailing list (LKML), R-help, R-devel, Wikimedia-l, and GNOME, system architecture design ideas are debated, code is written, and bugs ironed out. Since resource sharing is one of the central purposes of these lists, how distributed collaboration functions—its barriers and opportunities—is a persistent topic for investigation. Researchers are keen to discover more about the social structure of these lists: who contributes, the motivations of participants, and how leadership for project management is configured. Quite often, for example, the "eighty-twenty rule" will prevail. This principle holds that in collaborative project development "a relatively small number of participants shoulder most of the weight."[28] Testing this rule, however, one study found that a complex pattern of communication operates across the GNOME mailing lists, so that although a dominant set of contributors drive the discussion, a "casual group" may also engage productively.[29] Similar projects have investigated the social protocols in play on software developer mailing lists where it seems developers dominate certain threads, but also general list participants can figure in the range and frequency of list traffic.[30]

Volume of messages is not the only measure of a group's social structure and dynamics. Listening and lurking have been identified as valuable participatory strategies across online forums.[31] In one study, researchers examined group awareness within distributed software development channels. They found that mailing lists are often used by participants in ways that mirror in-person office arrangements because the "public nature of the mailing lists, and the possibility of overhearing conversations, is reminiscent of a co-present work setting—of several people sitting at different desks in an office and asking questions, stating comments, and making remarks to the entire room."[32] Analyzing the impact of mailing lists for the OSS community must also take into account the way conversations are dispersed across multiple platforms. Newer forums for knowledge sharing such as GitHub and StackOverflow are moving into the territory once dominated by mailing lists as the space for collaborative software development.

But it's not quite over for email groups. In their work on digital communities, Amelia Johns and Abbas Rattani show how the email list can create vital spaces of "pluralistic democracy" and citizenship for those often left outside formal decision-making processes or mainstream representations. In their digital ethnography of Mipsterz (short for Muslim Hipsters)—a "platform for young Muslims to share interests, collaborate artistically and discuss ideas," geared in large part for those who want to resist stereotypical views of Muslim women and youth—the authors note that although the Mipsterz community does utilize Twitter and Facebook, the email list acts as the "primary space where members seek to understand themselves: through cultural production, dialogue, and experimenting with ideas and thoughtful questions aimed at demystifying monolithic understandings of Islam."[33]

In both the American and Australian surveys, I explored current impacts of mailing lists and groups. People reported belonging to a diverse range of personal and business forums online. While some of these are work-related, many participants described their email groups as more "community minded" than work focused, comprising categories of special interest, hobbies, and support. Topics included "freecycle" swap groups, family history research, car-related and motor repairs, political parties, local council issues, sports, model railway, and health-related support resources. One survey respondent admitted belonging to "far too many to list," which amounted to more than 17 Google Groups and 10 Yahoo! Groups together with some privately organized email lists. A number of participants mentioned Yahoo! Groups as figuring highly within their repertoire of email practice. In many respects, Yahoo! Groups have flown under the radar of discussion platforms yet actually represent a richly inhabited conversation space. In 2008, there were approximately 113 million users, 9 million Yahoo! Groups, with availability in 22 languages.[34] Dating back to the late 1990s, Yahoo! Groups is often described as a "hybrid platform" of online interaction because it combines web-based forum with mailing list functionality. So participants can choose to communicate via online message boards or via email.

Many of the Yahoo! Groups, particularly those devoted to recycling like Freecycle, Freegle, and Full Circles, boast over 20,000 members, while others dealing with specific, localized topics such as regional math education may have less than 50 subscribers. Of particular interest to me is the highly active support network offering communication platforms across a vast range of health and wellness topics including resources for drug addiction, mental health issues, and family crisis. I spoke with Jurintha Fallon, who is the administrator of the Yahoo! Group called "The Other Half of Asperger's Syndrome," aimed at female partners and spouses of "individuals suspected to have, or officially diagnosed with" this developmental condition. The group invites people to "connect here with other women facing similar situations, who understand the realities of relationships with a partner on the spectrum." Explaining the strength of the group, Jurintha points to the way it creates a place where people feel safe to share their experiences of coping with Asperger's by offering "validation and understanding to the members that they are not alone." Reading across these groups I was struck with their storytelling capacity, both of the platform itself and of its users. Though many posts were brief—a couple of sentences to ask for specific recommendations of a service—just as many were told in narrative form. As Jurintha hints, the strength of narrative identification, finding someone with a story like yours, is a powerful tool for support.

List Moderation

In order to address questions raised so far about the contemporary significance of mailing lists and their social structures of governance and moderation, I now discuss what moderators said about how they manage their mailing lists and email groups. With a hunch that moderation functions as a largely invisible workforce, I was gratified to hear from a participant with substantial experience managing online public groups that he is rarely asked about list maintenance but that he often feels this work gets overlooked. Many of the respondents, however, shied away from defining this set of practices as "labor." Quite often group administrators or list owners would correct me by qualifying the work component of their duties, noting that "I just get rid of spam," or "it's only subscribing new members, that's all." But this shouldn't imply administrators aren't emotionally invested in their role. One mailing list moderator characterized their position as a "custodian" of the list, adding later they were "more the gardener, than the proprietor." For another moderator, Jasmine, accepting responsibility to the community requires diplomacy. She recounts a distressing situation where some of her Yahoo! Parenting Group attacked one of the members so she had felt duty-bound to intervene with a show of support for the besieged subscriber.

Conducting activities that seem difficult to characterize purely as work, whether that reflects the lack of monetary compensation or more broadly the absence of formally agreed-upon structures of workplace exchange, is what some have called "immaterial labor." During the late 1990s, commentators started to notice that the viability and sustainability of these new online platforms depended in large part on a community of volunteers. Making money for the pre–social media sector, content moderators and creators remained materially outside the benefits of the business model but crucial to its success. As Tiziana Terranova put it in 2000, regarding the emerging spaces of social interaction:

> Simultaneously voluntarily given and unwaged, enjoyed and exploited, free labor on the Net includes the activity of building Web sites, modifying software packages, reading and participating in mailing lists, and building virtual spaces on MUDs and MOOs. Far from being an "unreal," empty space, the Internet is animated by cultural and technical labor through and through, a continuous production of value that is completely immanent to the flows of the network society at large.[35]

Central to Terranova's critique was the legal action that volunteer moderators of America Online (AOL) took against the company for unpaid wages. At the time AOL was, as Hector Postigo describes, "a darling of the internet dot com bubble." With a controlling interest in Time Warner and over 25 million global subscribers, AOL maintained the "largest internal web community of any portal or Internet service provider."[36] Recognizing the worth of its vast social asset, AOL had functioned like a "Web 2.0" platform "a decade before the World Wide Web embraced the idea." For Postigo, "AOL knew how to exploit user participation before YouTube or MySpace."[37] Powering its message boards, online tutorials, and chat rooms was a 14,000-strong volunteer workforce called "community leaders" who would receive credits and discounts against their monthly internet access bills in return for their duties. The services they provided included forum content moderation and enforcing AOL's terms of use, but they were also "co-creators," a labor force responsible for "the creation of community, connectivity and social interaction, which became central components of the AOL experience."[38]

Although the relationship between AOL and its volunteers initially appeared to be benefiting all concerned, it began to sour when the company underwent a substantial restructure resulting in new pricing arrangements that no longer adequately compensated its informal staffing network.[39] Many felt they were being treated as unpaid employees rather than real volunteers because they were required to undertake training, submit timecards, and prepare reports. Meanwhile, AOL benefited enormously from the interactions generated by these community leaders, with one report estimating that chat alone produced at least $7 million a month. A class action lawsuit was launched in 1999, followed by a lengthy investigation by the US Department of Labor culminating in a $15 million settlement against AOL in 2008.[40]

Critical responses to this event have varied. As mentioned, Terranova, and others inspired by the Autonomous Marxist movement, talk about a "social factory" where capital is generated by activities outside of its traditional revenue streams, such as leisure time, thus constituting free or immaterial labor. Other terms to describe the global shifts in the logics of capital include creative labor, cognitive labor, affective labor, and precarity, concepts that are not simply reducible to each other but convey a growing awareness of the transformations occurring to working life in the twenty-first century.[41]

Drawing on this body of literature, Postigo finds fruitful avenues for understanding the AOL experience and by extension the interplay between contemporary digital platforms and their inhabitants. Yet he wonders whether this analytical lens brings certain limitations. As he explains, "questions remain over whether the conditions of these productive free labor communities can be understood solely from a perspective that sees their relations to capital as another form of capitalist exploitation of media consumers."[42] Similarly, in his investigation of the 2015 "moderator blackout" of news aggregator Reddit, J. Nathan Matias argues for a broader understanding of what he calls the "civic labor" of online moderation. Because moderators themselves respond simultaneously to the needs of distinct stakeholder groups—platform operators, community participants, and other moderators—critical perspectives must be similarly responsive.[43]

Illuminating accounts have also emerged from feminism. In her "feminist critique of the social factory," Kylie Jarrett points out that the domestic sphere has for decades provided surplus value for capitalist economics. Often dismissed by and overlooked in "the expanding explorations of immaterial labour and its economic role," the figure of the "digital housewife" shows how the supposed novelty of contemporary labor relations is untenable. As she astutely observes: "In my field of Media Studies, it often seems as if immaterial labour was only 'invented' when it moved out of the kitchen and onto the Internet."[44] Understanding contemporary patterns of "consumer labour" entails recognizing how the social contribution to capitalism, often made by women, has always been elided. As Jarrett explains:

> [The] history of domestic work also begins to reveal the centrality of immaterial labour to the capitalist system, not least because its differentiation from that which is deemed productive and compensated is integral to establishing a properly capitalist mode of production.... The privileging of strictly monetised exchange in capitalist systems and in critical thought has downgraded the importance of non-monetised production, viewing such activity as mere exchanges of fuzzy well-being that, by virtue of being beyond measure of the legitimised instruments, are not central to economic nor political calculation.[45]

Conducting research about the digital campaigns of feminist activism in Australia, Jessamy Gleeson looks at how social media moderators provide a source of free labor. She also queries whether theories of "exploitation" provide the whole picture. Gleeson's study concerned collectives that mobilize to protest sexist representations of women in the mainstream media, often perpetuated by "shock jocks."

Responding to these events, advocacy groups use Facebook and Twitter to target particular organizations by calling for product boycotts, sanctions against presenters, and stronger regulatory solutions. Running the social media accounts for these activist initiatives, moderators revealed that their volunteer workloads can cause burnout and stress due to the abusive content posted online and the harassment they endured; such hostile reactions often appear to be an almost inevitable consequence of feminist movements. Faced with these emotional demands, campaigners reported having to leave their group, a move that would put into jeopardy the ultimate success of a particular initiative. For Gleeson, these are the forms of digital and affective labor that need to be recognized if social media is to remain a viable tool for political action. As she argues, if "we continue to rely on the unpaid, digital labour of feminist protesters, we must also investigate how these same women can be best supported and their work sustained."[46] One of the interesting findings from Gleeson's interviews with these women's rights advocates was that terms such as "exploitation" must become nuanced in order to make sense of the online practices of moderation. Such complexity resonates with the responses received to my study of email group moderators, to which I now turn.

At this point it might be worth a reminder about the range of practices and software pertaining to the email groups considered. Mailing lists and email groups span both professional and personal interests and can be organized through informal networks or commercial platforms. These conversation spaces are usually constituted by automated mailing programs such as Listserv or via a provider like Yahoo!, with the latter offering a hybrid version of list connectivity because users can elect either the web interface or their email client to engage in a topic thread. I draw attention to such distinctions because the technical elements of group management figure highly in the tasks moderators need to perform. As a list manager on a mental health Yahoo! Group puts it, "One of the most challenging things that I have had with the group is figuring out the technological requirements for doing something that I don't usually do."

The final point of definition to cover is the designation of moderator itself. Across the literature, the terms "list owner," "administrator," and "moderator" are often used interchangeably. This is not necessarily a problem to resolve. Considering their use for three decades of online activity it's not surprising that these terms appear in diverse contexts, although some work has been conducted to produce typologies of moderation.[47] While moderators themselves use a variety of terms to describe their duties, for certain platforms these delineations will refer to specific socio-material configurations. On some forums the term moderation means messages are vetted before being distributed while on others it refers to a specific act of intervention performed by a person given that privilege. Such nuances are echoed across most networking platforms. In Facebook groups, for example, administrators and moderators have different levels of functionality. To flesh out the points made so far, I categorize email group governance into three categories: affective management, perceptions of labor, and the future of lists.

Affective Management

By this phrase I gesture toward how moderators are often required to help the group by creating a space for constructive debate and discussion. "Stepping in when there are disagreements that spiral out of control" is how CeCe More puts it discussing her duties moderating a Yahoo! Group about genetic genealogy.[48] As she elaborates, "Occasionally, there will be a difference of opinion in the group and it will turn into a very distracting, extended discussion." In those cases, she says, "I step in and call an 'end of thread' and remind the members of the group rules." Managing affect also refers to those moments where the work may drain the moderator's own emotional reserves. A pithy example comes from David Inouye, who founded and continues to run the highly successful list ECOLOG, a resource for the Ecological Society of America with over 17,000 subscribers.[49] When I asked him to reflect on a challenging moment in the list's history, Inouye sent me the retort from a member who disagreed with his suggestion to post a particular announcement to a more suitable forum, the name of which Inouye supplied. Unfortunately, the subscriber held a different view:

> You really are a prick, but I guess that comes with being a tenured prof. Yes, and I sent it there—but there are DIFFERENT people on each list. Also, if you were much of a biologist at all (clearly not, again characteristic of tenured faculty these days), you would know that 1) not everyone is on every list and 2) many labs do different things in different fields.... It is always logical and good practice to post this kind of request in as many places as possible—you might miss the one person with the best answer by not doing so.... So—Just pass it on! My next message may be a vote of the list to select a new moderator. Then, I may go outside and throw a rock, likely hitting a superior nominee. Do you actually think before you send these snide little messages to me while irrationally blocking them?

It turns out the moderators frequently receive emails sent "off list" remonstrating about particular posting decisions. And sometimes these mailing list disagreements can reappear in other "real life" spaces, as Gil Rodman discovered. Rodman is the founder and moderator of CULTSTUD-L, a list for the discussion of cultural studies–related topics including course syllabi, contemporary issues, and job announcements; it was established in 1995 and has a current user base of about 2,500. He recounts a particularly chilling example when a disgruntled poster actually tracked him down at home and phoned him to continue complaining about what was perceived as Rodman's egregious censure. It was also a case of mistaken identity. Apparently, the subscriber had thought the list was a forum for ex-members of cults—an understandable assumption, perhaps, given the contraction used in the list description.

Throughout their history, mailing lists have often sparked fears of censorship, and concerns abound on the silencing of "free speech." For long-term list administrators, the user base is now sufficiently educated as to appreciate that interactions must remain focused on the main topic of the list. Discussing his work with nettime,[50] the arts and activist network established by Geert Lovink and Pit Schultz in 1995, Felix Stalder explained:

> In the early days of moderation, there were lots of heated debates about "moderation = censorship," which can be nasty, but at the moment, most people seem to appreciate the somewhat bounded and focused nature of the exchanges. The flip-side of being forced to pay more attention is, of course, having to devote the time to do so.

Yet, newer lists and email groups also find themselves facing the same questions as subscribers grapple with the sometimes mysterious and invisible moderation processes and administrators try to cut noise and spam while allowing productive conversations to flourish. Or a slightly different situation can occur with moderators who are new to the role and trying to "get the hang of what to let through." It seems safe to conclude, then, that the trope of "moderation as censorship" is not a definitive discursive moment, but it will reoccur in both the life cycle of a particular list and within the learning curve of an administrator.

But what about when people are *too* quiet? Affective management is as much about encouraging dialogue as it is patrolling those hyper-expressive boundaries. Email groups often depend on the alacrity of a moderator to generate the exchanges. For Nikolas Lloyd, the administrator of Crossfire, a Yahoo! Group for miniature war gaming, time needs to be devoted to helping group members to contribute.[51] As he says, "The main thing I do is reply to lots of the posts, keeping threads interesting." Reflecting on how often he has to call a halt to difficult conversations, Lloyd says it's actually the opposite: "If anything, I encourage people to use the list more. Some people are frightened that they might say something too ignorant or irrelevant. People don't want to annoy anyone and can be a bit tentative."

Finally, the impact of the software and platform affordances should be taken into account when taking stock of the affective dimension of list governance. So often moderators are at the receiving end of subscribers' frustration about the vagaries of email program protocols. One list manager nominated "unsub" emails posted to the entire group as the least pleasurable aspect of his duties while another wished that subscribers wouldn't blame her for the limitations of the mailing programs.

Perceptions of Digital Labor

As I've already mentioned, notions of labor are tricky to canvass with those that engage online. This means that "exploitation" seems too blunt an instrument to capture what's at stake in group moderation, and to appreciate how its complex set of practices might unfold. Although administrators routinely dismiss the suggestion that their role involves significant work, equally they will then go on to enumerate quite substantial collections of tasks. Below is a summary of moderator responsibilities most frequently cited by my respondents, which I organized in descending order.

- Approve membership applications.
- Moderate all emails before being sent to group.
- Respond to emails with questions about how to subscribe, how to submit messages, and how to unsubscribe.
- Manage spam and ensure filters are set up adequately.
- Repost items sent to moderator but intended for the entire list.
- Discourage offensive behavior; make sure posts do not contain personal attacks.
- Maintain blacklists for known hostile subscribers and trolls.
- Block nuisance subscribers.
- Help members change their Group profiles on the website.
- Update and maintain the associated website.
- Initiate interesting debates.
- Talk to people about adapting their posts, make suggestions, and help rephrase.
- Reformat messages to improve readability and visual style; delete unnecessary quotes of previous emails.
- Make sure someone hasn't accidentally sent a private message to the list by mistake; intercept them and advise the addressee.
- Liaise with host server or IT department over technical issues.
- Write a daily column for the group, summarizing main points of discussion.
- Provide personal help and support privately, off list when asked.
- Intervene with controversial posts and explain to the community how the group rules help avoid tension and maintain a good feeling.

While certain tasks may appear to overlap, these actually entail slightly differing activities. If a list or group is unmoderated, then emails are sent directly from a subscriber to the group. But "hostile" posters may emerge at certain points and become a problem. In that case a decision can be made to change the software settings so the list moves to "moderation" mode and all posts are automatically sent to the list owner. Arriving at that decision necessitates a conversation between moderators, perhaps in liaison with a "leadership" or "executive" group that deliberates on such matters.

I was also interested to find out how people became moderators. Not surprisingly there are many diverse arrangements in operation, covering the spectrum from informal ad hoc systems to formal rules-based protocols, to appoint administrators of email groups and lists. One Yahoo! Group manager felt the process to be quite casual, explaining that his appointment came out of the blue: "The previous administrator got sick of it and assigned it to me. Without asking. Overnight. I woke up as the administrator. I didn't mind. Somebody has to do it." In some situations, the moderator is the person who has set up the group and, as mentioned, a number of these have been in existence for many years. VICTORIA, an electronic forum to exchange information on wide-ranging topics that bear on study of the Victorian era, is more than 20 years old.[52] Its founder, Leary, observes how this longevity plays an intimate role for its subscribers:

> The list has been a daily part of their working lives all of that time, and a medium through which they have made important and often lasting connections and collaborations. We've had deaths on the list, and births. We've even had romance—people who met via the list and later got married.

Other moderators might volunteer although this process can prove rocky. Having been a long-time and highly engaged member of her New York–based Yahoo! Parenting Group, Jasmine was excited to volunteer when Sarah, its current administrator, grew tired of the work involved and issued an expression of interest to the group. This was actually the second time in a year that moderation had to be sought. A year earlier one parent volunteered for the role only to anger the group by using her new position to change significant policy of the group. In place had been a process by which members could send emails designated as a commercial post (CP) to advertise childcare products or services. So as not to deluge the group with advertising, there was a weekly limit imposed of one CP per member. To the dismay of the group this new moderator announced she would introduce a fee for this previously free group feature. As Jasmine explained, "So many members reacted. It triggered several hundred messages, all in essence to support letting [the Group] continue to function the way it always did." After the outcry, the short-term moderator stepped down, which meant Sarah had to take back the reins. Once she had established accord and stability within the group, a year later Sarah again called for volunteers. Her children were now high school age and the time commitment demanded by the list was difficult to justify. This time the transition was smooth; Jasmine volunteered and assumed the role of moderator.

Within this particular scenario pulse many strands of labor. Looking at this story, no one could say that managing the delicate balance of the group and maintaining its overall sustainability does not require work. Yet each of these steps taken individually—the emails exchanged behind the scenes, the explanatory emails sent to the group, the catch-ups over coffee to work out strategy—may often seem slight: banal and trivial actions not worthy of the "labor" banner. But on Yahoo! Groups and discussion lists these micro acts of labor occur every day, helping to stitch together these spaces for interaction and support.

The Future of Lists

What then is the fate of email groups and mailing lists? Although it appears that many organizations are now choosing Facebook to host their group communications, email does remain preferable for others. And as with email more generally, the demise of the mailing list has been predicted for decades. As a number of list moderators note, when blogs emerged speculation was rife that these would bring to an end the use of email groups for discussion. Of course, it is very rarely an "all or nothing" decision. Groups, just like individuals, combine a range of choices across the digital landscape for their daily media consumption practice, forming what Nick Couldry calls "grids of habit."[53]

Some moderators report that their organizations will typically have a presence across both an email list and a social media account. In these cases, and referring back to the earlier discussions of labor, it's interesting to note that the role of moderator is becoming increasingly formalized. Both AoIR and Narrative-L appoint people to coordinate their digital communication over multiple platforms. For others a structure of volunteerism is crucial to what makes lists a better choice over commercial platforms.

Unbound by the imperative to monetize communication, some moderators feel that mailing lists offer an alternative to what gets called the "walled garden." This phrase refers to telecommunication, entertainment, or social media providers who attempt to restrict users from leaving their sites for products and services. Companies such as AOL in the 1990s, and more recently Apple, Google, and Facebook, have all been described as using this business model. For Stalder of nettime, mailing lists represent a little bastion of noncommercial activity; respite against the ever-hungry reach of the market:

> Being able to provide a forum that is totally non-commercial (volunteer labor, donated resources), that does not do any metrics or otherwise tries to "optimize" or automate the environment, is becoming ever more relevant as people understand better the substantial draw-backs of fully commercialized social media.

Other respondents—including Bryan Davis, one of the moderators for Wikimedia Labs, an initiative of the Wikimedia Foundation—echo similar comments. In helping to facilitate the work of the Wikimedia organization, Labs-L connects developers, supports project management, and provides resource sharing. Davis explains that the question of choosing the best communication platform comes up regularly on the list. Recently the group had considered a switch to a collaborative environment such as Phabricator or the open source team chat application Zulip. But lists retain the advantage in providing a durable resource for distributed conversations when policy intervention is required. Explaining those choices, he notes:

> I'm open to experimenting with alternate platforms, but ultimately for me personally it's hard to beat the openness of using email for conversations that should involve a highly distributed group of people on discussions. If policy decisions come out of conversations then I think they should be taken to wikis or other content management systems, but having the original discussion archives available for questions of "legislative intent" is very valuable in a distributed project. These are the kind of things that are overlooked in some arguments for replacement with either wall-garden solutions or chat-stream-based platforms.

Members of email forums with a specific focus on parenting and health mentioned that Yahoo! Groups offered a more intimate space than the big brands of social media. Some of the groups I got to know emphasized how a more localized nature helps deal with issues pertaining to neighborhood communities like school funding, caregiving facilities, or borough infrastructure development. While Facebook groups can, of course, provide an environment to support local community growth, they were somehow perceived as more public, less contained or bounded than the email groups, and hence less intimate. Asked to consider some of the differences between Yahoo! Groups and Facebook, Sirin Samman, the administrator of a parenting group called Inwood Kids, explained that its Facebook presence didn't seem to engender the same kinds of interaction. Because on Yahoo, she notes, "People are posting personal messages.... It just makes it feel like a more 'personal place,' [so] it is an extension of the neighborhood, but not visible to the whole city/world." Shaping this sense of a domestic intimacy was the frequent sociotechnical description of email as "coming straight to you," of "not having to go somewhere else to find the information." In the words of another moderator, an "online forum that insists the member go to the website to see the goodies doesn't work for people like me. I'm lazy. I want it to come to me." For every one of these commenters however, there were others who saw email groups and lists as being surpassed by newer platforms. CeCe More contrasts the 4,300 members of her DNA Newbie Yahoo! Group with the 50K members for the Facebook group that began only in 2015.

Conclusions

Email forums, discussion lists and groups have been in existence since the early days of the internet. In this chapter I had two main aims: first I wanted to establish the continuing vibrancy of email discussion lists in the era of social media and second to look at the role of moderator, particularly from the perspective of digital labor. In order to set up the discussion I needed to define some terms and parameters. So I outlined the formal properties of different types of discussion spaces that can be constituted through email. Here I argued why we can't simply gloss over the varying ways that the term "email list" is understood. Although phrases such as mailing list, newsletter, discussion forum, listserv, and email group get used interchangeably, they can actually describe quite different socio-material methods of communication. Perhaps at the end user level, variations are not apparent; but for moderators the software and cultural norms operating within a group can have significant impacts.

Definitions are also at play in the spam wars. While not all use of commercial mailing lists will constitute spam, what does often depends on your position in the email provision sector. As we saw in chapter 3, there is little agreement in the email industry about what consumers want from email. Marketers are convinced it's their message; email-hosting services like Fastmail aren't quite as sure. Perhaps when it comes to charity providers we should be a little more lenient, or at least recognize that email advertising strategies are often adopted by the not-for-profit sector or those interested in citizen-led, participatory democracy. Increasingly, community groups are looking at opens, deliverability rates, and other metrics specifically developed by email marketing in order to drive charity fundraising. More specifically, nonprofits may find themselves adopting some of the more aggressive strategies of email marketing to avoid the spam folder, such as using a deceptive abbreviation (Re: or Fwd:) in the subject line to make the message come across as a reply or forward.

Generally, these email lists and newsletters are transactional and so do not generate dialogue. In contrast the email discussion lists I then went on to discuss play a significant communicative role across professional and personal contexts. While research on Facebook groups is highly visible, less attention has been paid to the quite energetic and diverse forums of health and psychological support, professional development, resource sharing, and collegiality offered by mailing lists and Yahoo! Groups. Operating across differing levels of formality, from list software enabled to manually comprised "reply all" lists, this mode of communication constitutes a significant area of email practice. More precisely, I found that email lists provide a particular form of digital storytelling. In some of the support groups I explored, there was a strong sense of narrative as an empathetic tool to provide support. Even those lists that are generally formal or work-based in remit will often be sites where anecdotes are told in order to offer professional development or advice for inexperienced job applicants.

Behind the scenes for many of these groups is the work of governance. Group moderators and administrators perform fundamental duties, often almost invisible to subscribers, such as helping people with the technical elements of subscription, steering discussion in productive ways and maintaining the boundaries of civil discourse, or "ensuring members aren't bashing or attacking each other," as one moderator put it to me. When I asked about their role, however, people occupying these positions were often reluctant to categorize the activities they carry out to manage online sociality as work. Typically, respondents would dismiss the term "labor" or "work" only to itemize, often in quite a lot of detail, the many time-consuming tasks they undertook. Such responses are in line with current research about digital labor in content moderation and curation across online platforms where critical notions of "exploitation" remain contentious and problematic to apply. Nonetheless, it's clear that hours of thought go into carrying out online group governance, and I have tried to show why these micro moments of digital labor need to be factored in when discussing how email groups function.

Along with affective management and perceptions of labor, I was interested in what the moderators of email discussion groups had to say about the future of lists. In response, the point was frequently made that in the face of the overt commerciality of Facebook, email groups can represent a more intimate, informal, and less corporatized alternative for online social interaction. Never one to mince words, Geert Lovink calls social media "disastrous" for citizen engagement and public debate because of its limited graphic user interface and data privacy concerns. Alongside nettime, Lovink, with Korinna Patelis, has also established the Unlike Us initiative, a research network of artists, designers, activists, and scholars that is served by an energetic mailing list. Lovink sees promising alternatives being developed to challenge the walled garden of social media, including hybrids of existing mailing list cultures. One such hybrid is the arts and curatorial publishing platform e-flux and its discussion arm launched in 2014, named "e-flux conversations," which combines features of social media and the archiving software of mailing lists.[54]

This chapter marks the close of part II about affect and labor. Although the most salient point about mailing lists for my purposes is how moderation operates, my focus in this chapter on mailing lists as public discussions foreshadows "publics and archives," the subject of part III.

III Archives and Publics

6 The Enron Database and Hillary Clinton's Emails

During a turbulent university restructure, I mistakenly emailed an attachment to the dean. It contained a private plan that a group of us were proposing to senior management to form a new academic discipline. The discipline had no role for the dean. Thinking about this, years later, I still break into a clammy sweat. And of course, I am not alone. Here's a confession from one participant in my survey whom I asked, "Have you ever sent an email you later regretted?"

> I once sent a shameful email in which I called a business associate terrible and demeaning things. The email was meant for someone in my office but I hit reply-all instead of Forward. This was in the fairly early days of ubiquitous work email, say around 1998–99. I was so mortified upon realising my error that I ran to the server computer and considered pulling all the cables out, and asked the tech guys how to stop it, but it was too late. Since this gaffe many years ago I never put anything remotely controversial in emails and always quadruple-check the "to" line as habit, to the point of paranoia.

In this chapter I examine the various ways that email can surface—especially the technical and social conditions that allow emails to be read beyond their intended audiences. The emergence of vast, publicly accessible email datasets such as the "Enron email corpus" offer rich opportunities for research but also raise questions of ethics. How do we think through what is at stake when private communications become public? What are the policy and legal implications of the ever-increasing volumes of metadata? More broadly, since the availability of scalable archives locates email at the interstices of the public and private spheres, how does it help us to track some of the practices through which these domains are constituted? Finally, architectures and structures of data give way to content. What are people saying in these emails? Reading my survey responses I was fascinated to discover the stories that people recounted about the varied paths of their emails: how the sociotechnical mistakes and slips in the course of an email exchange opened up unintended audiences.

Email Boundary Crossing: Public, Personal, Professional, and Private Domains

To open the final section of the book, I look in this chapter at the publics formed by email communication, together with the archival infrastructures that support it. I use two case studies to explore how email can travel across multiple domains, whether as the result of hacking or inadequate forms of information security. On other occasions the unexpected release of email can come from a mistaken keystroke, as when emails are forwarded without the knowledge of their original authors or when people are inadvertently copied into an address line. Still other examples might involve formal processes where documents are made public to satisfy the legal requirements of discovery. Behind all these cases there are people writing mundane emails, sending notes home about dinner or arranging catch-up drinks with colleagues. And this is what interests me most about the two case studies I selected. Although they touch upon big questions of email security, they also reveal the everyday nature of email. So, before I get to a discussion of the Enron corpus and Hillary Clinton's email server, I look briefly at some different approaches for understanding how emails are able to travel across public and private domains.

Trying to come to grips with privacy in the abstract is not very helpful. Instead, it's more worthwhile, as many commentators note, to examine how privacy works in highly relational, subjective, and situated contexts. According to Christena Nippert-Eng, "Privacy is one ideal-typical endpoint of a continuum; 'publicity' is at the other end." And yet, she explains:

> As with all ideal-types, this is only an analytical tool, useful to think with and help make sense of the world around us. Real-world experiences, real-world things, fall somewhere in between these two analytical end-points. Something may be relatively more private or relatively more public, but it is never purely either.[1]

In this sense, notions of privacy control, management, or selection afford a more productive way to frame the issue of how people exercise agency over what is revealed and what is concealed. Nippert-Eng applies the metaphor "islands of privacy" to areas traditionally protected from trespassing, such as wallets and purses or doorbells and windows, to underscore the important role of boundaries in gauging how privacy is attained or breached. Email too is considered an "island of privacy," particularly as it relates to "social accessibility." The degree to which people are able to control their availability to others via email is often directly connected to questions of hierarchy; for example, people in positions of power seem to have more say about whether an email needs an immediate reply. In previous chapters we have already looked at workplace interruptions, "reachability," and email overload; here I am more interested to highlight Nippert-Eng's insight into the boundary work of email management and power because it frequently plays out in the two case studies I discuss below. In particular, I will be exploring the unexpected release of personal and professional emails into the public domain.

If boundary crossing is one of the ways to think about email privacy, as Nippert-Eng explains, other areas involve email's own technical security features; these may be inbuilt or provided by third-party plugins through processes of encryption, verification, and authentication. In the early days of the internet while protocols were being developed, email was mainly exchanged between users known to one another and there existed a fairly high degree of trust about the origin of a message. The aggressive commercial use of email, through spam or phishing, had yet to be conceived of by those working on the early systems for email transport.[2] As one 2014 report puts it, "Limitations in technical design, security provisions and legal solutions render email an 'ideal' attack vector for a variety of cybercrimes."[3] In order to combat security threats, the US-based Online Trust Alliance (OTA), an initiative of the Internet Society, recommends the use of email authentication software. Authentication, they explain "helps create a feedback loop between legitimate email senders and receivers to make impersonation more difficult for phishers trying to send fraudulent email."[4] Whereas authentication software usually happens at the enterprise level and is not necessarily visible to end users, two-step verification methods require more work from email consumers who need to enter passwords to receive codes, usually via mobile phones. The lack of such verification and authentication techniques was thought to be responsible for the massive Deloitte email breach in 2017, where hackers accessed the company's global email server and retrieved data from a reported 5 million email accounts.[5]

Leaks, Hacks, Breaches, and Releases

Since email gained widespread use in the late 1990s, leaks have been a feature of the communication landscape. Leaks, thefts, hacks, breaches, and releases—these terms all describe different social, technical, and legal practices. The so-called Sony Hack, as I go on to explain, differs markedly from, say, the release of the Panama Papers.

In 2014, the private emails and other confidential information of Sony Pictures were widely published, apparently by hackers calling themselves the Guardians of Peace. Journalists enthusiastically parsed this information, but it was very difficult to navigate until WikiLeaks produced an indexed and searchable version, "The Sony Archives," in April 2015. Julian Assange, the editor in chief of WikiLeaks, explained in a press release that publishing the archive served the public interest because it showed "the inner workings of an influential multinational corporation" and was also "newsworthy and at the centre of a geo-political conflict." As such, he claimed, "It belongs in the public domain."[6]

Although releasing 173,000 emails to illustrate the machinations of the US film industry sent international ripples across the sector, the Panama Papers operated at a much grander geopolitical scale. Almost 5 million emails were accessed by an anonymous German whistleblower and made available to the International Consortium of Investigative Journalists (ICIJ) in 2016. These documents revealed the financial arrangements of the Panama-based law firm Mossack Fonseca, which was responsible for creating offshore entities, namely tax havens and trusts in more than 200 countries and territories, dating back some 40 years. Implicated in the disclosure of these often illegal financial dealings were 140 politicians worldwide, including the prime minister of Iceland; the king of Saudi Arabia; the then prime minister of the UK, David Cameron; and Russian president Vladimir Putin.[7] Even though the ICIJ spent a year analyzing the information and published their findings, the actual database is not easy to access. Explaining its rationale for selective publication, the ICIJ states it will not "disclose bank accounts, email exchanges and financial transactions contained in the documents."[8] During an interview with *Wired*, the director of ICIJ, Gerard Ryle, defended the decision not to release the full database to the public by arguing, "We're not WikiLeaks. We're trying to show that journalism can be done responsibly."[9] WikiLeaks deplored the decision with a tweet posted in April 2016 stating "DC based @ICIJorg is setting a very dangerous & short-sighted international standard where everything is censored by default."[10] In 2017, the ICIJ added the "Paradise Papers' to its database, a cache of documents leaked from the offshore legal provider Appleby, showing international networks of tax evasion.[11] Three years after the Panama Papers were released, the ICIJ has estimated the tax revenue reclaimed internationally as a direct result of these leaks stands at more than $1.2 billion.[12] *The Laundromat*, a film based on the Panama Papers and directed by Steven Soderbergh and starring Meryl Streep, Gary Oldman, and Antonio Banderas, premiered worldwide in September 2019.

WikiLeaks is a key repository for many of these email releases and has become a touchstone for broad discussions about information transparency, legacy journalism, nationhood, and the public construction of knowledge.[13] In addition to the releases mentioned, the WikiLeaks site is host to vast caches of indexed email databases. Among them is the so-called Global Intelligence Files, a corpus of over 5 million leaked emails released in 2012 from the private intelligence company Strategic Forecasting (Stratfor) covering the period July 2004 to December 2011. Revelations from this dataset include apparent evidence of widespread insider trading, drug trafficking, and corrupt political deals either occurring within the company or between significant political actors. Whether these claims can be verified—and this has certainly been questioned—what remains salient to me is the public accessibility of emails written with the expectation of some degree of confidentiality. Other notable email releases orchestrated or hosted by WikiLeaks include more than 44,000 emails and 17,000 attachments of the Democratic National Committee, leading to the resignation of its chair, Debbie Wasserman Schultz, in July 2016, together with the related Podesta emails consisting of the hacked inbox contents of John Podesta, Hillary Clinton's 2016 campaign chairman; and the Macron campaign emails, a searchable archive published in 2017 comprising 71,848 emails and 26,506 attachments of the French president.[14]

But WikiLeaks is not the only source or archive of leaked emails. Moreover, those email files sitting tantalizingly on the WikiLeaks servers are often surprisingly difficult to navigate. In some cases, very limited data-mining features are available, meaning, for example, that subject lines cannot be searched. In these situations, other media outlets are often relied upon to parse and translate the plethora of data.[15] And even when the content can be fully accessed, and there's no denying these caches are substantial, the stories behind the disclosures—the paths traveled by the data, the cultural and legal constraints in operation—are sometimes rendered invisible by the drama of their ultimate release.

Case Studies: The Enron Database and Hillary Clinton's Emails

In order to unearth some of these pathways, in this chapter I consider in detail two occasions when private emails have been released into the public domain, each event operating under different legal and social constraints. The first concerns the internet publication of internal emails from the energy company Enron by the US Federal Energy Regulatory Commission (FERC) in March 2003. This action was taken during the FERC investigation into the fraudulent activities of Enron that resulted in criminal findings against the corporation and jail sentences for a number of its key executives. The dataset has breached legal protocols of privacy but at the same time has been used extensively by computer and social scientists for a diverse range of research projects and new methodologies across data science, linguistics, and management theory. While these studies have enriched the critical terrain, very little has been said about the people who became subjects of this research. I was curious to know more about what happens when someone's emails are used as public datasets. Wanting to give voice to the subjects of big data research, I spoke to a number of former Enron employees about their experiences of personal and professional emails made public. Almost 20 years later, what has it been like to live in the archive?

I then examine how the US Department of State released Hillary Clinton's private emails written during her time as US secretary of state. The publication was the result of legal actions under Freedom of Information Act legislation, together with Clinton's own request, following a story in the *New York Times* breaking the news that she had used a private server for official business in her role as secretary of state. As I explained in the introduction, the methods for these case studies differ. While it was difficult and time consuming to track down former Enron staff who were willing to talk about their experiences, it turned out to be completely impossible to apply the same method to the Clinton emails. All my approaches went unanswered. Rather than talking to people, then, what I illuminate with the Clinton case study is a contemporary example of how everyday work practices can have quite dramatic consequences.

Exploring each of these instances opens up email in everyday contexts: email published in exceptional circumstances, yes, but emails composed during the routine banalities of work and domestic life. We should not, however, too sharply separate the intimate minutiae from its institutional settings. Indeed, as I have been trying to show throughout the book, bureaucracy and its media forms—emails and their bloated attachments or spreadsheets and their fury-inducing incomprehensibility—is a significant force in structuring affective labor.

The Enron Email Corpus

Before its spectacular collapse in 2001 Enron was one of the world's most successful energy corporations. Shortly before declaring bankruptcy it was named no. 7 on the Fortune 500 list, its shares peaked at $90.75, and its CEO, Ken Lay, was paid $67.4 million in one year.[16] As the shares plummeted and stories of corporate malfeasance emerged, nearly a quarter of its workforce, some 4,000 people, lost their jobs and life savings.[17] Founded in 1985 as a gas and electricity provider, Enron of the 1990s was hailed as an innovative, knowledge economy firm in ways that resonate with the rhetoric about disruptive technologies and contemporary start-ups. Creating new commodity markets into broadcast audiences and internet bandwidth, its "maverick" and "freewheeling" brand of entrepreneurship reflected its dot com era.[18] The downfall of Enron, how it became synonymous with corporate corruption, was driven by a byzantine financial reporting system that gave an inaccurate picture of capital and risk by understating its liabilities and overstating its equity and earnings.[19] In addition, the increasingly deregulated energy market meant that Enron could manipulate supply, a fact that came to light during the California power crisis of 2001, where the company created electricity blackouts in order to drive up the price of power. At the time, California experienced crippling and, it was later revealed, artificial power outages costing the public billions in surcharges.[20] Phone call transcripts released for the investigation show how a cavalier Enron official persuaded an employee of the Las Vegas power plant to find an excuse to blackout the grid: "Ah, we want you guys to get a little creative, and come up with a reason to go down … anything you want to do over there? Any cleaning, anything like that?" The employee at the power plant helpfully responded, "OK, so we're just coming down for some maintenance, like a forced outage type thing?" And the Enron staffer then gave approval: "I think that's a good plan.... I knew I could count on you."[21]

At least as troubling as the illegal accounting procedures was the workplace culture of Enron. Terms such as "brutal," "punishing," and "excessive" abound in the mythology that shapes its stories. The *Wall Street Journal* blamed the corporation's bosses for creating a culture of "pushing limits."[22] Again, these damning yet oddly admiring descriptions can't help but remind us about the history of innovative or vanguard business models and their utter dependency on particular narratives about how workplaces should be fashioned. The relentless efficiency demanded of its employees by Google, Amazon, and Uber is matched only by the power of these myths to circulate alluringly through our media landscapes. As was the case with reactions to "the right to disconnect" law discussed in the introduction, a work culture of constant availability is often highly praised because it displays the grit, determination, and entrepreneurship of its employees.

Jeff Skilling, one of Enron's top executives, apparently embodied the punishing ethos of the company through the team-building exercises that he led. These were adventure trips staged across the world, to places like the glaciers of Patagonia and the Australian bush. During these "daredevil expeditions" employees were challenged to engage in risky sports activities including trail bike riding through inhospitable terrain. One trip, a 1,200-mile road race, saw all its participants injured; another had the goal of finding an adventure "where someone could actually get killed."[23] Appearing before a congressional committee, Sherron Watkins, a vice president at Enron, spoke of its "culture of intimidation," an assessment shared by many others in the company.[24] As misgivings grew about corrupt business practices, Watkins felt powerless to intervene or to communicate these fears up the Enron management chain, describing such a course of action as a "job-terminating move."[25] Indeed, so widespread and evocative are these tales of workplace harassment that Enron has become one of the most widely cited case studies for the ethics of employment relations and management theory. And being able to track the role of email has been central to these studies. As Eric Gilbert explains in his work on organizational hierarchy, the "Enron email corpus is without parallel in the research community. Nowhere else can you find such a rich, complex, and naturally occurring email dataset."[26]

Going Public

So how did the private correspondence of Enron employees become public? Shortly after Enron declared bankruptcy in December 2001, the FERC launched an investigation to gauge the "potential manipulation of electric and natural gas prices" involved in the West Coast power supply markets. Part of their remit was to demand that Enron secure and make available "any books, papers, correspondence, memoranda, contracts, agreements" that would reveal whether the company had engaged in these price-fixing activities.[27] The problem was that by this time, as Joe Bartling explains, Enron was no longer operating as a functioning corporation. Their infrastructures and systems were fast being dismantled, and 80 percent of their workforce had been sacked as the company collapsed. The company did not have the resources, largely IT personnel, to comply with FERC demands of discovery. Instead, the FERC used the third-party contractor Aspen Systems Corporation to access and store the Personal Storage Table (PST) files of Enron's Outlook accounts. Bartling was responsible for archiving the PST files to CD format and uploading to web servers with parsing Concordance software for linguistic analysis.[28] Following the submission of the FERC report finding, that Enron had engaged in creating artificial volatility and price distortion, in 2003 the email corpus was declared by the FERC to be in the public interest; it subsequently released online and in hard drive format 1.6 million emails exchanged between 150 Enron staff from the period 2000 to 2002.[29] The majority of the sample contains emails exchanged between the top executives Ken Lay, Skilling, Watkins, and Andrew Fastow, but it also scooped up the messages of ordinary Enron employees, their family members, and their friends. Such complexity, diversity, and reach deliver to us a poignant resource while simultaneously issuing a number of methodological challenges.

Data Mining Projects Using the Enron Corpus

To give a picture of how the dataset has benefited the research community, I want to take a look at some of the projects utilizing the corpus and the research methods these have developed. As might be expected, managing the database has not been easy. For one thing the sheer scale has made retrieval and access a problem. In fact, this situation itself has provided computer scientists with resources to use in the development and trial of new applications for data retrieval, search, analysis, and visualization. A major contribution in this regard is the data science project "Cognitive Assistant that Learns and Organizes" (CALO) based at the California research institute SRI International. This DARPA-funded initiative, which resulted in significant achievements such as the development of Apple's Siri program, was responsible for the initial processing and cleaning of the raw data extracted from the Enron servers. Leslie Kaelbling of MIT purchased the data files for $10,000 and worked with Melinda Gervasio at CALO to remove spam, duplications, and attachments in order to provide clean sources of searchable archive material for the research community; the result is generally referred to as the CALO dataset.[30] Yet this shouldn't imply that the Enron email corpus exists in a pristine, discrete, stable form; part of the business of its analysis has involved reformulating it for specific purposes. Bryan Klimt and Yiming Yang, who cleaned the dataset to test the functioning of threads and filtering within an email system to classify and organize communication, conducted one of the earliest studies.[31] Jitesh Shetty and Jafar Adibi, who built a MySQL database from the original corpus to provide a test bed for fraud detection and counterterrorism research, applied another early approach.[32]

For William Styler, the overwhelming volume of machine-generated material including email headers, signatures, automated leave responses, and requoted material has significantly hampered the linguistic potential of the sample. The problem with these duplications arises because they give a false sense of the naturally occurring rhythms of language, he explains:

> If an otherwise vanishingly rare word occurs in an important email from an important person, the number of messages which technically contain that word (by virtue of quoting) will make that word seem much more common, even if it was seldom used intentionally in original.[33]

Reasoning that people tend not to forward spam or other automatically generated texts means that the sent mail folders can promise some degree of integrity. Styler then reconfigured the set to make it suitable for language parsing tools and created the EnronSent Corpus. This resource has now been used across a wide range of language-based projects including machine learning algorithms, social media sentiment analysis, and mobile interface design.[34] Where Styler was focused on discourse, other efforts to redesign the datasets have been directed at Social Network Analysis (SNA). Large-scale communication databases obviously lend themselves to SNA methods because they reveal the connections made between people.

In these studies, it is imperative that relations between interlocutors be accurately mapped; a central problem researchers faced with the Enron corpus was that multiple email addresses existed for the same person. In many cases this would not affect the actual delivery of an email, but it would introduce "noise" into the analysis. Such mapping concerns were further compounded by the fact that employees themselves would, as is often the case in any organization, be worried about whether they'd sent the email to the "correct" address, and they'd resend certain messages in order to resolve possible errors. As Yingjie Zhou explains, "When the people who send and receive emails are of interest in SNA research, mislabeling a person may lead to confusing or even wrong conclusions." His PhD work implements strategies for cleaning archived organizational email datasets by consolidating email aliases and thereby offering a revised version of the Enron emails for SNA research purposes.[35]

A similar research project was launched in 2005 at the International Conference on Data Mining run by the Society for Industrial and Applied Mathematics (SIAM), which produced two themed issues of the journal *Computational & Mathematical Organization Theory* devoted to methods for analyzing the Enron corpus. Two key questions emerged in the analyses: What is the inter-organizational profile of a firm as it moves toward crisis? How does message traffic change over a corporate lifetime? Again, prior to addressing these topics was the need to remove ambiguity from the data and a number of these studies designed new software applications for this purpose. An interest in organizational crisis, for example, motivated Jana Diesner, Terrill Frantz, and Kathleen Carley to enhance the sample by building functionality to better map email addresses to people and by contributing fuller details of position descriptions and titles. Regarding the latter axes, the authors wanted to correct for conflicting and multiple job titles in use across the dataset and thus devised a "career history" for each of the cohort.[36] As the editors of these journal articles observed, technological issues of method are at least as significant as the patterns of social relations revealed by the Enron corpus.[37]

In exploring the vast array of research projects that utilize the Enron corpus, I have limited the field to those who reconfigured the actual archive, whether by developing new indexing tools or cleaning and reposting the dataset. So, it's worth briefly acknowledging that many significant research projects have also drawn on the Enron material to answer questions across disciplines from social sciences to digital forensics studying topics such as the hierarchy of workplace email,[38] the so-called gendered use of email,[39] surveillance technologies,[40] and social network analysis.[41]

Ethics

If technical challenges arise, ethical questions have also shaped the archive. In response to complaints received by FERC from former Enron employers, the original database was further refined and the corpus edited to remove identifying information including social security numbers, bank account details, employee performance evaluations, and photographs not deemed relevant to the investigation. In a series of reviews implemented during March 2003 to February 2004, the FERC considered requests concerning 124,697 email-related documents and subsequently withdrew thousands of items from the dataset.[42] Yet of course, ethical problems persist and continue to be addressed by social science researchers and data mining software developers. A figure central to many of these discussions is William Cohen of Carnegie Mellon University who first made the corpus available to academics and who maintains one of the key sites used for downloading the dataset. Here Cohen urges researchers to be "sensitive to the privacy of the people involved" pointing out that "many of these people were certainly not involved in any of the actions which precipitated the investigation" and are not therefore subjects of legal discovery. Cohen himself has coordinated the editing of a number of versions; he explains that his policy is to "remove emails from this collection when requested by the email's authors or recipients," and he provides an annotated history of which emails have been removed.[43]

The sense of shock at the sudden, and for some entirely unjustified, disclosure of private information is shared by a number of the former Enron employees I interviewed. One of the respondents currently working as Divisional Operations Director for a UK energy provider describes his indignation:

> I couldn't believe that Enron had provided the emails to the government.... There had been no attempt by anyone at Enron to identify and withhold purely personal, non-work related information. Apart from the impropriety of releasing employee's personal information, I couldn't see why Enron would release confidential business information that was unrelated to the government investigations.... I worked for the gas pipeline division. To this day, no evidence has emerged that the gas pipeline division engaged in any misconduct. Nonetheless, thousands of confidential emails relating to the gas pipeline division's business were publicly released. Competitors, customers, and anyone else could access

those. Whoever made the decision to release the emails had acted in a completely irresponsible manner.

As a way to deal with both technical and ethical dimensions, the Enron Data Reconstruction Project (EDRC) was launched in 2009 by the information governance consultancy Electronic Data Recovery Model (EDRM). This standards organization brings together law firms, consumer groups, government bodies, information retrieval providers, and researchers working in the legal discovery and data preservation markets. EDRM provides products and services for the e-discovery sector including privacy guidelines, statutory compliance advice, and datasets for testing and prototyping methods of information retrieval. Responding to many of the scalability issues faced by the original CALO dataset, the EDRC project aims to make available reconfigured Enron datasets, in particular one that rebuilds the original corpus to a near native format of PST files with attachments. In 2012 the EDRC published new versions of the Enron emails with the Amazon Web Services Public Data Sets, a repository of large-scale databases including the Google Books Ngrams corpora, the 1000 Genomes Project, and the Wikipedia Traffic Statistics.[44] This Amazon version was updated in 2013 when EDRM produced a resource that was "cleansed of private, health and financial information" since it had been an "open secret" that the database contained personal data of former Enron employees. Despite the earlier efforts from FERC and CALO to clean the corpus of personally identifying information, 10,000 items required removal including credit card numbers, social security information, birth dates, home addresses, and a wide variety of medical and legal material.[45] Discussing their methodology, Matthew Westwood-Hill and Ady Cassidy note that companies face considerable financial and legal risk when internal data retention systems are not robust. It was clear from the documents inspected that employees regularly duplicated information outside the Enron firewall; using flash drives or email and cloud storage systems, "convenience copies" of spreadsheet attachments, for example, might be sent by staff members to their own personal email account for storage and backup.[46]

The publication of this now-scrubbed dataset provoked widespread debate. There were two strands of discussion: First, a number of commentators in the legal and digital forensic community wondered how this raw dataset had been utilized as a research tool for so long in a format that revealed personally identifying information. A second but related critique concerned the still-faulty dataset that EDRC had failed to clean. On this latter point,

Jim McGann of Index Engines, a data security and information-management company, demanded, "Why facilitate and publish a data breach?" Reviewing the revised dataset, McGann argues that substantial disclosures remain: the "republished document is still littered with many social security numbers, legal documents, and other information that should not be made public." Running a search on "social security number," Index Engines discovered substantial data breaches "buried deep in email attachments, sent folders and Outlook notes."[47] Through the Twitter hashtag #NuixChat, the EDRC project responded to such criticism by establishing avenues for revising the database by bringing to their attention further privacy violation issues.

Living in the Archive

Identity theft and other potential consequences of these data leaks and disclosures have received attention, but what it might mean more generally for a person whose correspondence exists in the Enron email corpus has been much less discussed. To my knowledge, no one has gathered together the participants in the dataset to find out how the publication of their emails has affected their lives. So in June 2016 I attempted to do just that.

As I immediately discovered, people are very reluctant to discuss their experiences of the emails going public. Of the 135 former Enron staff I contacted, most did not want to participate; many responded politely that their lives had "moved on" and said they had no wish to dredge up what had been, in general, a painful and regrettable time. A couple of participants actually had no idea that the database existed as an extensively mined information resource, and in fact my initial contact had prompted their searching for themselves, often with feelings of trepidation. "This is the first I've heard of anything being public," replied one, adding "you nearly gave me a heart attack to think my emails are out there for anyone to read!" As I discussed in chapter 4, this unforeseen and thorny issue suddenly confronted me head on. In the course of a project about the ethics of database research, I had now thrust people, perhaps unwittingly, into the heart of the debate. For every one of these replies, however, there were those who actively welcomed the opportunity to discuss their experiences. One participant noted that no one had thought to ask permission from former employees to release the emails in the first place, or considered how it might have affected their lives during the past 15 years.

If my initial contact with these former Enron employees came out of the blue, the announcement that their emails were to be released by the company was also a surprise arriving, as it did, through ad hoc and informal channels. As one participant explained:

> I first learned from a coworker that the emails had been made public. She said something along the lines of: "Oh my God! They've posted all of our emails on the internet!" At first, I thought that she had to be mistaken. Of course, I was aware that the government was conducting an investigation and that Enron had made certain relevant emails (and other documents) available to the government as part of that investigation. But I worked for the natural gas pipelines division of Enron, which had nothing to do with the accounting irregularities and other misconduct that was being widely reported in the media. And I wasn't involved at any time in any of the matters under investigation. So, I couldn't see how my emails (or those of the vast majority or my coworkers in that division) would have been provided to the government.
>
> Shortly thereafter (I think the same day), another coworker provided a link to the internet site where the emails were posted. There was a search feature that allowed anyone to type in an employee's name and determine whether any of their emails had been posted (or whether their names appeared in any email strings). The text of each email was available for anyone to read. So, I ran a search using my last name and was shocked to learn that a substantial number of emails (perhaps all) that I had sent or received since I joined Enron roughly two years previously had been posted.

Although most were unaware of an official process of notification, there was a general sense of what was happening, that the emails had been posted to the government website.

> I think that someone in management let us know when they were provided, as we were told to review our emails and identify those related to the topics requested for the lawyers, and then someone told us they had been released in general, and someone else mentioned they were going to be provided to a site where they could be searched a little later.

Reports in the media had alerted a number of Enron staff to the release; one participant said that having heard, he then quickly performed a search to verify what specific emails might be "out there." This was a reaction shared by other participants; understandably people wanted to know what specific emails of theirs had been published. Thinking back on their initial responses to the data release, people noted that today's environment is very different to the way it was then:

> My reaction was one of surprise but also a little annoyed that this was in public domain, only because I had always thought this was internal work product. Nowadays I would have had a different reaction, knowing how companies and others track emails.

> The email I saw that had my name ... was fairly innocuous. But I have seen some emails that discussed career progression and were not exactly friendly. It reinforced in me to not write things that I would not be comfortable to see in the public space. I also felt a bit sorry for the people that were talked about in the email in an unfavorable way.

For others, the salient aspect of this initial release was the company's cavalier attitude to the email privacy of their employees, above all for those who had no dealings with the main figures of the investigation:

> I couldn't believe that Enron had provided the emails to the government. A few of my emails were personal (I was generally careful not to use my Enron email account for personal correspondence). Like almost everyone, I had a handful of emails that contained personal information such as my bank account and social security numbers. And there were emails to or from family members and friends. Those emails were all posted. There had been no attempt by anyone at Enron to identify and withhold purely personal, non-work related information. A number of coworkers had emails that contained very detailed information about their personal lives. One friend was going through a bitter divorce and his emails contained communications with his wife and others that should never have been disclosed to anyone.

Participants felt ambivalent about searching the database themselves once they learned of its existence. Some did admit to having had a certain curiosity over the years: "My first reaction was 'lol, these will be interesting! I must have a look.' And in fact I might go and try to find them again now for another look." Others commented that the tools in place made navigation of the corpus quite difficult. As Brian noted:

> When I first found out, I checked out the dataset myself to see if it was actually true. I mainly was curious which of my emails had been posted. I found a few, but really the dataset was quite difficult to search. I wasn't particularly interested in reading my coworkers' emails and didn't feel right doing that.

A general sense of disquiet pulsed through the recollections when it came to reading the emails of coworkers, one that was difficult to disentangle from the lasting negative impression the company had made on its staff:

> I looked at some of them, but focused primarily on my own. ... I was mostly curious what kind of "stuff" was "out in the wild," so to speak (especially mine). I did

> read some of the emails from others, but got bored fairly quickly. Since my experience with Enron wasn't entirely pleasant, I didn't want to relive it; so I decided it would be best if I just let it go.
>
> I didn't seek to ... read the emails of coworkers—it felt like voyeurism and an invasion of privacy. The whole thing felt like a witch hunt at the time, and there were plenty of people at Enron who were doing their job in good faith who were hit by the fall out.

Interestingly, however, as a former project manager at Enron Online discovered, there was in fact an upside to having his emails out in the public domain over the years. As he remarked, "Honestly being on that list of emails in several instances has been a boon." He added, "Folks look up that dataset and realize that I have worked at Enron. Today that can be an advantage as most people in Houston and the US energy business recognize that some of the folks who worked there were some of the best in the business."

What I learned from speaking to these former employees at Enron is that the everyday work environment of email was central to how they viewed the publication. Although a number noted that "transparency" and "compliance" were important for any business, and that any inherent problems of an organization should be made public, the fact that little consideration was given to those caught up in the drama was an equally pressing issue. Now I turn to another instance where personal emails became public in the context of work.

Hillary Clinton's Emails

In March 2015, the *New York Times* revealed that Hillary Clinton had been using a private, unsecured email server to host official communications exchanged in the course of her duties as then US secretary of state.[48] This story came to light as a result of several events. First, it emerged as part of the House Select Committee on Benghazi, an investigation into the September 2012 attack on the American embassy in Libya that killed the US ambassador and foreign service officer. The existence of a private email account, the so called "home-brew server," had also drawn attention because of the activities of a hacker known as Guccifer, who had apparently accessed the accounts of Clinton's colleagues in 2013. Finally, use of the personal address had attracted interest because of an increasing volume of Freedom of Information applications lodged by media outlets and political organizations.

Just one month after the *New York Times* exposé, Clinton announced her intention to run for the Democratic nomination in the 2016 US presidential race. But the email problem was to plague her entire campaign. Following the election of Donald Trump as president of the United States in November 2016, Clinton's private email server has become a shorthand explanation for his success. Certainly, the popular Twitter hashtag and meme "but her emails" continues to express strong opinions about the role played by the server and the weight given to it by media outlets since that time.[49]

In April 2019, the findings of a 22-month FBI investigation into allegations of conspiracy, collusion, and Russian involvement in the election were released to the public in redacted form. Titled "Report on the Investigation into Russian Interference in the 2016 Presidential Election" (or the Mueller Report), it found that the "Russian government interfered in the 2016 presidential election in sweeping and systematic fashion."[50] While there is much to explore here in terms of democracy, journalism, and media discourse,[51] I am taking a slightly different interpretation as I trace the path to publication of Hillary Clinton's emails.

The Server

When Clinton assumed the role of secretary of state in 2009, the server was set up at her home through a commercial IT provider, Platte River Networks, using the domain names clintonemail.com, wjcoffice.com, and presidentclinton.com, which pointed to her email address hdr22@clintonemail.com. Between 2009 and 2013 Clinton used her private email account exclusively, apparently never registering for the official state "dot gov" address. Not only did Clinton use this personal account, a number of her aides were also given access to the domain "clintonemail.com" to set up their own email addresses, which were then used for official communication purposes. Indeed, being issued with one of these Clinton email addresses conferred a certain amount of cultural cache on its owner.[52] According to the National Archives and Records Administration (NARA), the US government agency with statutory responsibility for collecting and maintaining government records, utilizing a personal email account to conduct state business contravenes regulatory and conventional email practice. During her tenure as secretary of state, from January 2009 to February 2013, government regulations operating under Section 1236.22, which oversees the management of electronic mail records, stated that federal bodies who allow their employees to "send and receive official electronic mail messages using a system not operated by the agency" must "ensure that federal records sent or received on such systems are preserved in the appropriate agency record-keeping system."[53]

In one sense, this clause can be interpreted as permitting the use of private email servers by government agencies, and Clinton herself certainly maintained this line. Defending her actions a FAQ sheet released by Clinton's 2016 presidential campaign team explained that in keeping with this stipulation, Clinton's practice was to email "officials on their '.gov' accounts, so her work emails were immediately captured and preserved."[54] However, many political, administrative and industry figures have not agreed with this interpretation. The NARA explained that the "National Archives discourages the use of private email accounts to conduct Federal business, but understands that there are situations where such use does occur."[55] Further clarification came from the former director of litigation at the NARA, Jason Baron, who elaborated in an interview with the Senate Judiciary Committee:

> Any employee's decision to conduct *all* e-mail correspondence through a private e-mail network, using a non-.gov address, is inconsistent with long-established policies and practices under the Federal Records Act and NARA regulations governing all federal agencies.[56]

Not surprisingly, Clinton's political opponents were quick to attack. In most cases the target of their ire was lack of transparency and the contravention of national legal and policy frameworks on information management. Ari Fleischer, a former Bush aide, challenged Clinton's explanation that it was simply easier to use her own account and devices: "Personal convenience? Hah. She did it because she only trusts a few top aides and wanted total control," he posted on Twitter.[57] Reince Priebus, the chair of the Republican National Committee, accused her of risking national security, making secret promises to foreign governments, and "hiding her emails from the administration, journalists, and the American people."[58] Fox News argued that Clinton had topped the list of "worst ethics violators of 2015."[59] And the right-wing extremist media outlet *Breitbart* declared, "It's a good thing laws apply only to little people in Obama's America," before going on to say:

> It all has such a marvelous banana-republic feel to it, which should outrage the taxpayers forced to pay trillions of dollars for a government that can't even follow the sort of document-handling requirements it routinely imposes on small business owners.[60]

But criticism also emerged from the left side of politics. One Democratic senator commented that although Clinton was not breaking any laws, the silence over her motivations for using a private email account was hurting her reputation.[61] It was also reported that key campaign donors were withdrawing support,[62] while political commentators opined that "Democrats are right to be angry" because Clinton had attracted unnecessary scandal through an avoidable mistake.[63]

Following the *New York Times* article the NARA wrote to the US Department of State expressing concern about the "potential alienation of Federal email records created or received by former Secretary of State." It further requested an explanation of the record management policies:

> If Federal records have been alienated, please describe all measure the Department has taken, or expects to take, to retrieve the alienated records. Please also include a description of all safeguards established to prevent records alienation incidents from happening in the future.[64]

In response, the State Department explained that policy reviews were occurring even before the news broke in the media. In 2014, for example, it had tightened up the policy on personal use of email and general record keeping, instructing its senior staff not to use Gmail or other private email providers for "official business." It also expressly forbade staff from deleting personal emails. "At no time during designated senior officials' tenure," it stated, "will their e-mail accounts be cleared, deleted, or wiped for any reason." Although senior officials are permitted to delete personal emails, they should bear in mind that the "definition of a personal e-mail is very narrow. The only e-mails that are personal are those that do not relate to or affect the transaction of Government business."[65] Together with issuing internal directives and clarifying policy in 2014, the State Department had also requested that former secretaries of state—Clinton, Condoleezza Rice, Colin Powell, and Madeleine Albright—make available any "federal record such as an email sent or received on a personal email account" during their tenure if there is "reason to believe that it may not otherwise be preserved in the Department's recordkeeping system."[66]

As a result of this letter, in December 2014 Clinton's office provided 55,000 pages of printed email correspondence, or approximately 30,000 emails, to the State Department. Some of this material was subsequently tabled in the Benghazi investigation. In her cover letter Clinton's chief of staff, Cheryl Mills, assured the State Department that all statutory obligations had been met but that "out of an abundance of caution" the emails were being made available, noting many of these were "not Federal" but personal and should be handled accordingly.[67] Interestingly, the other former secretaries of state did not supply their email records. Apparently, Secretary Rice did not use a personal account, and although Secretary Powell did use a personal account, he had deleted the emails. Meanwhile, it appears that Secretary Albright, who was in office from 1997 to 2001, never used email—either a US government account or a personal server—during her term.[68]

What Is a Work-Related Email?

Confusion swirls around timelines and motivations for the release of the Clinton emails. Different players clash in their versions of events, with some asserting the email publication resulted directly from demands made by the House Select Committee on Benghazi while others, like Clinton herself, play down international security instead emphasizing the bureaucratic aspects. Her office, she explained, responded to routine requests from the State Department to "help with their record-keeping" because of the "technical problems that they knew they had to deal with." She had complied with their requests, issued as she pointed out not just to her but "to previous secretaries of state," which "asked that we, all of us, go through our e-mails to determine what was work-related and to provide that for them."[69] But if her office had shared these particular emails, what had it withheld? Further questions were raised about the process by which emails were categorized as work or personal and, among the latter, what was their fate? From the original cache of 62,320 emails sent and received during the period 2009 to 2013, about half or 30,490 were determined as relating to official government business and subsequently supplied to the State Department. The remaining 31,830 were classified by her office as private and were apparently deleted from the server.[70] And here's where an interest in the materiality of communication really comes to the fore. To understand the process by which these emails became public requires an appreciation of the everyday labors of institutional life. Raking through the emails to classify their content relies on a substantial infrastructure comprising people and systems, and these processes have been enacted at two significant moments.

The first sweep occurred at Clinton's office in 2014 as lawyers prepared to release to the State Department emails of "federal record"; in other words, those emails that were deemed work related. According to Clinton's office, classifying the material involved a thorough process to "review each email" and would be "erring on the side of including anything that might be even potentially work-related." What this search actually amounted to has been the topic of spirited debate. Specifically in doubt was whether each of the emails was read individually or instead "reviewed" across the entire cache using automated search software. Fact sheets released by Clinton's office provide details about the search criteria but do not use the word "read" in relation to the emails. Clinton staff performed various keyword searches across both the header and body fields. Beginning with the address field, a search was conducted for any ".gov" suffix; it produced more than 27,500 emails and represented 90 percent of the email cache eventually submitted to the State Department. In order to capture any possible non ".gov" correspondence a search was then carried out for first and last names of specific State Department and other US government personnel, including "Deputy Secretaries, Under Secretaries, Assistant Secretaries and Ambassadors-at-Large" as well as "Special Representatives and Envoys, members of the Secretary's Foreign Policy Advisory Board," and, finally, a number of "senior officials to the Secretary, including close aides and staff."

Next the cache was sorted into sender and receiver categories in order to safeguard "against non-obvious or non-recognizable email addresses or misspellings or other idiosyncrasies." Lastly the cache was searched using keywords such as "Benghazi" and "Libya." These search steps, which focused on the non ".gov" fields, delivered a further 2,900 emails.[71] During a press conference held a couple of days after the news story, Clinton described the personal content and her reasons for withholding the information:

> At the end, I chose not to keep my private personal emails—emails about planning Chelsea's wedding or my mother's funeral arrangements, condolence notes to friends as well as yoga routines, family vacations, the other things you typically find in inboxes. No one wants their personal emails made public, and I think most people understand that and respect that privacy.[72]

For many commentators the retrieval system used for this filtering was decidedly underresourced and poorly designed. Seizing on the issue of whether each email was individually read, critics from both the left and right weighed in. *Slate* called the system "dodgy," pointing out that if aides or other officials also used nongovernment email addresses and did not happen to include the designated key terms, their emails would not make it to the public domain.[73] The leading conservative political blog, Hot Air, sneered at the claim that Clinton staff had read the individual emails—writing "er…sure they did"—and argued that the need to use automatic filters was negated if the emails were individually assessed.[74] In response to the media speculation Nick Merrill, press secretary to Clinton, clarified the initial information that had been released:

> In wanting the public to understand how robust of a search was conducted, the fact sheet laid out several examples of the methods used by the reviewers to double and triple check they were capturing everything. It was not meant to be taken as a list of every approach performed to ensure thoroughness. Those subsequent steps were in addition to reading them all, not in lieu of reading them all.…We simply took for granted that reading every single email came across as the most important, fundamental and exhaustive step that was performed. The fact sheet should have been clearer every email was read, which we are doing now.[75]

Interestingly, although the FAQs went through a number of revisions during 2015, these were not updated to include the specific term "read."

Clinton Emails Released

At this stage, however, the Clinton emails were still not publicly accessible. The second email sweep took place at the State Department in preparation for publishing the cache. Here is a brief summary of what such a sweep entails: First, identify those that do not count as Federal documents and are deemed personal. Next, remove or redact both "classified" and "sensitive" information. Classified information refers to issues of national security, diplomatic relations, or law enforcement and is divided into the subcategories of *confidential, secret, top secret,* and *top secret / sensitive compartmentalized.* Information deemed as classified is circulated through a separate email system that the State Department and its agencies use, which is "walled off from the Internet." It is illegal for classified information to be circulated through the official government email server, meaning that even if Clinton had used that system the State Department would still need to check documents for their classified status before releasing.[76]

The illegality is also an issue: in late 2015 the FBI launched criminal proceedings to determine whether Clinton or any of her staff circulated classified messages through unsecure channels.[77] Information regarded as "sensitive" includes personally identifying information such as bank accounts, home addresses, or social security information. Here again, reading is important. Where the initial assessment of the data in Clinton's office had relied substantially on machine-readable search processes, at this point people begin to actually read the emails. But reading potentially classified information requires clearance, and reports emerged that the State Department was unable to fulfill these personnel requirements. As the *Washington Times* explained, the "State Department is struggling to hire enough staff to review the huge load of emails" because the "staffers need to have top secret clearance."[78] In July 2016 the FBI announced it had not found evidence of any legal violations. But in late October, only days before the US election, James Comey, the then director of the FBI, reopened the investigation, which again did not result in criminal proceedings. As mentioned, many commentators, including Clinton herself, have viewed the fallout from the highly circulated media narratives of suspicion as central to her eventual defeat.[79]

Piles and Piles of Paper

Paying attention to the reading practices of these emails reminds us again of the media materialities operating at the site of their access. The initial cache was provided to the State Department as 55,000 pages of printouts collated into 12 "bankers" boxes measuring approximately 24 × 15 × 10¼ inches. Each box was labeled with details of the dates pertaining to the emails within.[80] Faced with such daunting paperwork, much has been made of how to comprehend its sheer scale. Let's visualize it: stacked one page on top of the other, the pile of printouts is "high enough to let someone climb up and change a light bulb 18 feet high," and would take the average person "more than a year to type" and "more than 76 days to read."[81] A corpus of paper is also, of course, difficult to search electronically, a fact germane to those who believe Clinton was hiding information about her campaign funding arrangements and possible conflicts of interest in the Clinton Foundation. Refuting such accusations, the Clinton FAQ sheet explains that the State Department requires hardcopy because the archival capacity of electronic storage is not deemed sufficiently robust.[82] And then there's the problem of getting all that paper online. Once the content has been cleared for release the documents need to be converted to the appropriate file formats. This scanning process consists of five steps that include both automated and manual aspects. Each document was scanned and then manually coded so that the To, From, CC, BCC, Date (sent), and Subject fields could be searchable. An automated process to detect duplicates was then applied, which often resulted in someone manually having to read and assess the possible duplication. This exacting process of converting paper to digital is rendered more complicated by the fact that some, but not all, of the paper records are double sided.[83]

The process of scanning the documents as PDFs and uploading them to the State Department Freedom of Information Act (FOIA) website was time consuming too, and caused delays in achieving the publication deadlines established by court order. The technical infrastructure and personnel resources required to vet, scan, convert, and upload the original paper documents has been estimated at a cost of over $1 million. Beginning in May 2015, monthly batches of emails were posted to the website with the final publication occurring on February 29, 2016.[84] When the December 2015 tranche was posted on New Year's Eve it was met with criticism about how the State Department was falling behind its schedule.[85] Complaints of delay have characterized the entire uploading process, and chief among these have been accusations of political maneuvering by the State Department, supposedly keen to shield Clinton and the Obama administration from scrutiny. Amid the presidential campaign, further FOIA lawsuits resulted in additional email releases by the State Department in late 2016, although many of these were thought to be duplicates.[86]

Although the political contours are impossible to ignore, I am more interested to trace the everyday institutional practices that shape the Clinton email corpus. I am struck by a gnawing sense of familiarity about the scenes describing the laborious journey these emails have taken on their way to the public domain. During a court case launched by Associated Press against the State Department about its delays in responding to Freedom of Information Act requests, the judge criticized procedures for processing a submission relating to the release of a subset of 60 emails. "Any person should be able to review that in one day—one day," the US district court judge Richard Leon remarked, caustically adding that "even the least ambitious bureaucrat could do this."[87] The target of his ire was the head of the FOIA office, John Hackett, who defended the delays by pointing out that funding cuts in the department meant he did not have the staffing resources to adequately meet the FOIA demands. With 60 full-time staffers working on the technical aspects of the FOIA submissions, sourcing the requisite security clearance for these email scrutineers meant it fell to a select group of Foreign Services officers who, it turns out, could only be employed on a part-time basis.[88]

Clinton herself echoed these frustrations regarding publication delays. In the days after the media story broke, she made several announcements about her use of the email server and her various plans for dealing with the ensuing issues. A tussle emerged between the timetables of the State Department and Clinton. On March 5, 2015, she posted to Twitter, "I want the public to see my email. I asked State to release them. They said they will review them for release as soon as possible."[89] As delays with the scanning process at the State Department became apparent Clinton grew more public about her displeasure at the protracted progress. During a press conference she answered questions about the emails and responded, "Nobody has a bigger interest in getting them released than I do," adding, "I respect the State Department, they have their processes, they do for everybody not just me, but anything they might do to expedite that process, I heartily support."[90] A couple of months later her criticism became more pointed as she urged the Obama administration to pick up the pace, explaining that the State Department are "the ones that are bearing the responsibility to sort through these thousands and thousands of emails and determining at what pace they can be released," again emphasizing, "I really hope it will be as quickly as possible."[91] In subsequent interviews and media releases Clinton reiterated the line that it was she who had instigated the document dump of emails. Calling the situation "an interesting insight into how the government operates," she pointed out, "if I had not asked for my emails all to be made public, none of this would have been in the public arena."[92] Many, however, would disagree. For one thing, the numerous Freedom of Information Act applications had been in process for at least two years before the *New York Times* story broke about her personal email accounts. And as noted earlier, the State Department itself had requested access to emails sent by a number of Secretaries of State. The political backdrop forms an important part of the context for understanding the Clinton emails, but I am again drawn to the prosaic nature of email practice. Let's now turn to the motivations for Clinton to adopt a personal account.

Convenience and the Reach of Shadow IT

Though the media materialities and bureaucratic frames for uploading the email corpus online pique my interest, I am also attentive to the everyday use of devices and software that frame the event. Why *did* Clinton use a private email server? Obviously there's a credible narrative about avoiding scrutiny, but what is less explored has been her explanation of "convenience." To me, the Clinton emails need to be understood within the prosaic flows of everyday work life and the recognition of an increasingly prominent IT shadow within organizations. In her initial explanations of the private server, Clinton blithely dismissed any nefarious undertones to her actions. As she put it, "When I got to work as secretary of state, I opted, for convenience, to use my personal email account.... I thought it would be easier to carry one device for my work and personal emails instead of two." The choice to use her private email, then, came from her not wanting to be weighed down. In retrospect she conceded, "It would have been better if I'd simply used a second e-mail account and carried a second phone, but at the time this didn't seem like an issue.'[93]

Unlike contemporary smartphones that provide apps for accessing multiple email addresses, during her tenure in 2009 the Blackberry model issued to government staff did not have a secure functionality to access different email accounts.[94] On Twitter, Jon Favreau, a former speech writer for Obama, defended Clinton's explanation by pointing out that White House (WH) and State Department devices did not allow two email addresses to be accessed from a single unit until 2011. Moreover, Gmail was not accessible via any of the White House computers.[95] As he explains, "It was such a pain. Eventually there was an app for your iPhone that allowed you to access your WH emails."[96] Other government employees also recount difficulties with using the official federal IT system, in relation to the issue of Bring Your Own Device (BYOD). According to Theresa Brown, again on Twitter, "Gov't was not BYOD friendly from 2005–11 when I was at DHS. No private email on gov device. Blocked webmail too."[97] Unfortunately for Clinton this was not a recollection shared by everyone. Working in the Obama administration at the same time as Clinton in 2009, Ray LaHood, secretary of transportation, maintains that he conducted both private and official business on a single Blackberry device.[98] More broadly, many commentators seized upon Clinton's subsequent revelations about her predilection for carrying multiple devices, rather than traveling light as she had insisted initially. In subsequent interviews she admitted to being "two steps short of a hoarder," regularly transporting a plethora of phones and devices including iPhone, iPod, iPad, and the infamous Blackberry.[99]

The point here is not to prove once and for all Clinton's real motivations, if that is indeed even possible, but to place the discussion within the increasing reliance on shadow, "stealth" or "rogue" IT across government and private sectors. These terms describe the situation where third-party applications or cloud services such as Dropbox, Evernote, WhatsApp, and Gmail—together with personal devices or other hardware—are used by people in workplaces often in direct contravention of the organization's stipulated record keeping or security policies. Although some of these applications are huge global brands and may not seem "shadowy" in any sense, their unregulated status within an organization means they can operate as an unacknowledged resource. It is their organizational noncompliance rather than their market success that confers the term. Reliance on shadow IT represents risk for data protection and information governance and is seen as a substantial challenge to organizational infrastructure globally. According to its 2018 "Shadow Data Report," Symantec finds that on average organizations will have over 1,500 cloud-based apps on their networks, most of which are adopted without official IT sanction or oversight.[100]

While many in the legal and IT communities were interested in the Clinton emails from this perspective—one called her "the poster child for the dangers of rogue IT"[101]—there was little recognition for the ways in which these practices function within broader diverse economic arrangements. In other words, as we have learned from research into the shadow industries of cinema and the internet, these alternative, unregulated, or informal cultural and economic practices often shore up mainstream systems of organizational communication.[102] Commercial and public-sector institutions alike may well decry the prevalence of shadow IT through acceptable use policies, but their business models often rely on the use of personal devices and the competencies of employees to access software and hardware outside those supplied. More generally, an institutional policy that demands staff use only official hardware and software applications is clearly at odds with a growing insistence on productivity and labor flows that operate beyond the standard working week. Again, while the political context of Clinton's email practice should obviously not be ignored, my interest in shadow IT in this case remains in the everyday push and pull between different forces within a workplace and the various motivations or media choices made.[103]

Public Access to the Clinton Dataset

Like the Enron emails, the Clinton dataset has been accessed and reconfigured by researchers and journalists. The *Wall Street Journal* (*WSJ*) substantially improved the access and search capacities on those provided by the Department of State FOIA website in its hosting of the corpus. For instance, the *WSJ* interface is impressive for annotating key personnel with bibliographic details including profile picture, together with providing search filters such as email date ranges and sender/receiver and subject lines. To this suite of analytics, it added an interactive tool for archival purposes. Announcing the project in May 2015, the *WSJ* urged readers, "Join the document dive! Read and tag Hillary Clinton's emails, and then submit your findings."[104] This feature, however, appears to have been limited to the first document release, the so-called Benghazi emails. In addition to these functions, the *WSJ* ran a live blog each month when new emails were published, parsing the content with key topics covered and giving contextual information relating to relevant legislation or applicable foreign policy.[105] Meanwhile, the data science crowdsourcing platform Kaggle has reformulated the original datasets as downloadable CSV files and a SQLite database, inviting members to contribute code and analytics. Predominantly using data visualization methods, participants have designed ways to scale the corpus to produce results about countries most mentioned, sentiments, wordclouds of subject lines, and volume of recipients.[106] Further studies have involved natural language processing skills and statistical analyses of the metadata to plot her "inner circle" and their emailing patterns.[107]

As an apt coda to this case study, in May 2019 the emails became part of the Venice Biennale for a solo exhibition called *HILLARY: The Hillary Clinton Emails* by the poet and artist Kenneth Goldsmith. To "create" the work, Goldsmith printed and displayed on colored paper 60,000 of the Clinton emails in an effort to demonstrate their quotidian and unspectacular nature: "an anti-monument to the folly of Trump's heinous smear campaign against Clinton."[108] Adding to the whimsy of the exhibition was the fact that Clinton herself visited the show and was photographed leafing through the emails.[109]

Conclusions

In answer to questions posed about their email practice and perceptions, my survey respondents wrote vividly of emails delivered wrongly, addressed in error, or unexpectedly made public. People were fascinated about what others might disclose in their messages and felt just as keenly the embarrassment when they themselves had revealed personal details not intended for wide dissemination. This got me thinking about how best to capture these experiences beyond the individual responses. I chose the two case studies because they each illustrate a different aspect of the story of disclosure, how everyday emails are made public. In turn these stories of disclosure have two distinct threads. First there is the content: what gets revealed within these vast, publicly accessible datasets about our communication practices. And second there is the narrative of how these emails became public: the legal, technical, and social paths traveled to their eventual publication.

Tracing these various trajectories throws light on our different understandings of what constitutes public and private email. Neither of these categories exists as an absolute, of course. As Nippert-Eng has shown, private and public domains always exist in tension, creating shorelines and boundaries across which communication is made. In addition, the terms to describe such situations—leaks, hacks, breaches—cannot be used interchangeably. For many working in the IT or legal sector, by its very nature email exchanged within an organization should be deemed as public communication to the extent that it is often archived by law for compliance or discovery purposes. But in practice many people write their emails with the expectation these will not to be made available to the public. Once emails have moved to the public domain questions will arise about the socio-material conditions leading to their release, inevitably framed around justification (whether the release was warranted, for example) while also attracting popular interest to see what might be exposed.

While vast datasets of email such as the one represented by the Enron corpus have proved indispensable to the research communities from within the fields of data science, machine learning, management theory, or linguistics, what goes less remarked is the impact felt by those who are the subjects of widespread and ongoing visibility. Sitting in conference rooms during the course of my email research, looking at Powerpoint slides displaying datasets carrying email addresses or graphs with all personnel helpfully revealed and annotated, I have been troubled by a persistent question: "Sure, your tables are pretty, but how do the actual people feel about the public exposure that made them possible? As is often the case with research, the answers weren't precisely what I expected. For one thing, I was unprepared for my own implication in the ongoing story of personal emails made public. As noted, some of the Enron participants had no idea until I contacted them that their emails were on the public record. When one of the participants revealed that I'd nearly given them "a heart attack" in telling them about their public emails I was, to say the least, taken aback. It had not occurred to me that they that were unaware, and that I would be the one to alert them to the publication. To balance my own sense of disquiet, I was then quite relieved to discover some former Enron staffers were actually gratified the emails had gone public and, moreover, had benefited professionally from their release. Enron also inevitably throws light on our current times. There is an increased awareness and perhaps acceptability of the diminishing borderlines of privacy. A number of Enron participants mentioned that what had seemed unthinkable in 2003—the widespread publication of an organization's internal emails—has now become almost an everyday occurrence.

Shifting focus to another government release occurring over a decade later, I was again intrigued by the publication of the Hillary Clinton emails. My concern was not the overtly politicized media coverage, although I do not mean to downplay the significant role played by email in the 2016 US presidential election and beyond, or to challenge Clinton's argument that convenience, not cover-up, was her motivation for using a private server. Rather, my purpose was to chart the material circuits of publication and the institutional frameworks through which these emails flowed—and to argue that the everyday contexts of workflow email often get underplayed in the interpretation of Clinton's private email server. Specifically, I tried to show that there are hidden stories revealed by the prevalence of shadow IT in the workplace. In diverse work settings, these third-party, open source, or cloud-based applications are being used to an increasing degree. So widespread is the growing reliance on this unregulated sector of the market that a budding industry has sprung up alongside the practice to coach organizations on how to avoid the inevitable security threats posed. The convenience and apparent cost efficiencies, however, make shadow IT, and the related BYOD, difficult to resist.

Across government offices, university lecture halls, company boardrooms, and nonprofit workshops people are busy accessing Gmail or Zoom as they go about their routine workdays. This might seem like a banal observation, but I suggest that it shows a crucial aspect of our everyday working lives. Speaking personally, I can attest to working with senior university staffers as we swapped files on Dropbox, only to hear them speak forcefully in faculty meetings to ban the practice and use the official network drive instead.

Finally, what we have seen in these two case studies is how the archive shapes email both for researchers testing a range of hypotheses from diverse disciplines and also for those who have "lived in the archive." In the next chapter I extend this observation by looking at how artists are imagining the archive in their practices where, unexpectedly perhaps, email provides an evocative subject matter.

7 The Art of Email

When we think about email, "art" is hardly the first thing that springs to mind; the routine nature of email, its sheer banality, surely cannot make our hearts sing. Not only is it mundane, it is supremely irritating. Email overload with its persistent invasive demand for our attention—how can people make art about such a thing? And yet they do. In this concluding chapter I want to explore how email has inspired thoughtful, funny, and provocative work—contemporary pieces that can be traced back through the elaborate signatures of the 1990s, to mid-twentieth-century mail art, and further back to eighteenth-century epistolary novels, the stories told in letters became a model that would provide the imaginative backdrop for a number of email projects.

If email is difficult to think about aesthetically, the relation between art and technology more generally is a path well traveled. The history of this discipline variously termed "media arts," "new media art," "digital art," "telematics," "image science," "bioart," or "data visualization" stretches back before the internet but seems especially to come into focus during the early 1990s. As with any exercise in categorization, the cluster of terms above describes practices with differing degrees of specificity; perhaps what distinguishes this form of aesthetic endeavor is the relative emphasis given to the chosen media. This means that debates within the field often turn on the prominence of the medium, its form, versus its content or theme. Thus, as Robert Mitchell explains, for example, "bioart" can capture both the media used—microorganisms, bacteria, cell lines, plants, insects, animals—or it can utilize the traditional tools of painting or photography, to explore relevant themes such as environmental degradation and genetic modification.[1]

The relation between art and technology harks back long before the twentieth century, with notable examples including Albrecht Dürer's use of the newly emerging printing press technology in the sixteenth century or the nineteenth-century interest in automata. Bioart does not necessarily mark the first instance where artists borrow the techniques of science and technology, as Mitchell points out: "one could probably make the argument that historically artists have often been among the first to exploit the capacities of new technologies and the implications of new forms of science."[2] But there is something about bioart in particular and media arts more generally—an interest in nontraditional materials and often controversial themes—that makes these fields a useful vehicle through which to ask fundamental questions about the purpose of art in our culture. To take the bioart example again, with its often-radical deployment of bacteria and human tissue as the preferred media, its practitioners have faced real legal threats and accusations of bioterrorism, prompting art critics to recommend it be banned from gallery spaces and curation. Such highly charged moments are opportunities for public discussions about what counts as art.[3] And as we shall see with an email example from data visualization, artists and scientists are collaborating in ways that blur the boundaries of both disciplines.

The archive has become a significant site for digital arts exploration. Historically, the collecting, housing, ordering, and display of cultural or natural artifacts have always been important if not coexistent with art practice. Emerging in mid-sixteenth-century Europe, the curiosity cabinet or *Wunderkammer* helped usher in the modern museum and shaped contemporary discourses of representation by establishing a critical link between classification and aesthetics.[4] For digital artists since the 1990s, archives and databases have provided resonant source material for explorations of both cultural and personal memory and the ways in which the body itself can be coded, organized, or digitized.[5] From the Human Genome Project to The September 11 Digital Archive, people have been drawn to archival technologies as the symbolic inspiration for creative work, the repository for those pieces and the media through which that creative work can happen. Perhaps the archive is such a rich site for artistic expression because the processes of storage, search, and display are almost instantly recognizable from our daily encounters with the world. As Gabriella Giannachi observes:

> Not only is information about us located, at any one point, in a plurality of archive, we also incessantly archive our lives, primarily but not exclusively, through social media, and use a plurality of archival materials to orient ourselves within our everyday lives.[6]

As we saw in the previous chapter, email can operate as a "boundary object," its public or private nature variously defined, or used for conflicting purposes, by different social groups.[7] For the IT manager an email system is private and should have appropriate encryption or authentication software in place. To the legal department, however, all email circulated through its system ought to be treated as public to the extent that it can be published under statutory powers of discovery. For the organization's employees, email might be written or conceived as semiprivate in recognition of a company IT policy, but it can still be shocking if the contents are disclosed. And finally, for the academic researcher, the archivability of email makes it infinitely accessible if not, as was the case with Enron, always ethical. Given its capacity to travel across public and private borders, to evoke both the familiarity of the everyday and the dramatic emotions of sudden disclosure, it is no surprise that creative practitioners have considered email a compelling topic.

In this final chapter, I bring together two aspects of email that percolate throughout the book: how email functions as a contested "object" poised on the brink of privacy and publicity; and, in turn, how email's history involves multiple encounters with the post office. From the late 1970s, when the USPS began to worry their business would be swallowed up by email, through to today's client software, which still displays the iconography of "mailboxes," the two communication systems have been in constant dialogue. So, in this chapter, I discover further links between them.

Epistolary Literature and Mail Art

Using the conceit of personal emails made public, writers and artists in the last few decades have been exploring the nature of private correspondence. During the mid-1990s as email moved from the province of computer scientists to become more ordinary and widespread, a flurry of literary activity resulted in the email novel. In quick succession came titles such as *Chat: A Cybernovel* (1996), *Virtual Love: A Novel* (1994), *Safe Sex: An E-mail Romance* (1997), *Email: A Love Story* (1996), *Single White E-mail* (1998), and *The Metaphysical Touch* (1999): books that fictionalized the so-called found emails between, usually, romantic partners.[8] While the content here was fiction, other writers have used the conceptual model to create works that dance around the edges of fiction and fact. One such work is *I'm Very Into You*, a 2015 book that reproduces the emails exchanged between the artist Kathy

Acker and the media theorist McKenzie Wark during a two-week period in 1995.[9] The two met in Sydney, embarking on a brief fling. These published emails document what happened next: a flirtatious, seductive, angry, gossipy, and scholarly epistolary relationship. It is a difficult and curious work to encounter, not least because of the perhaps unintended moments of vulnerability exposed. Writing in the preface, Matias Viegener, the editor and executor of Acker's estate, admits to moments of doubt about whether these emails should be published. Indeed, a famous unnamed novelist apparently declined the invitation to write the foreword when approached by Viegener and Wark because reading the emails "felt too much like rooting around in someone's underwear drawer."[10] Nonetheless it is a compelling example of the storytelling capacity of emails. Before looking at the creative use of email in more detail, I want to provide some historical context by examining its trajectory as an aesthetic expression.

Epistolary History

Just as the post office was becoming more efficient, opening up new paths of communication and connection, it suddenly began to figure across the literary landscape. People started writing about the postal service in their letters to each other, wondering about the precise time that a letter might arrive, sending apologies for its delay, speculating about its journey. Not only was the postal service an absorbing topic of interest between partners in real conversation, it became a rich source of fiction. The rise of the epistolary novel in the eighteenth century was enabled by a diverse set of social, economic, and material conditions including increased rates of literacy, industrial and commercial advances in book printing, the proliferation of letter writing manuals, the growing numbers of female writers and readers, and developments in the efficiency of the postal service.[11]

Throughout the nineteenth century the genre adapted, and although novels composed entirely of letters seemed less visible, the postal service itself was everywhere seen. Perhaps the quintessential text in this regard is Bram Stoker's *Dracula* published in 1897. The novel is often thought to lie outside the traditional epistolary genre, but its narrative structure is animated by a plethora of communications media including typography, shorthand, gramophone, telegrams, and letters. As a number of scholars have noted, nineteenth-century fiction persistently grapples with the cultural transformations wrought by the post office of the time. Faced with the sudden speed-up of daily communication, which made letters cheaper to post and quicker to receive, fiction responded by representing these material systems of support through its pages. What Laura Rotunno calls the "ubiquity of letters" in Victorian fiction is explained by the "rise of postal reform between 1840 and 1898 that inspired the lower and middle classes to read and write more frequently" and "allowed this population a measure of social and political agency." This meant more people entering "the literary world as writers and readers."[12]

Similarly, in his study *Telegraphic Realism*, Richard Menke charts the "information systems" of Victorian-era fiction, to show how that kind of writing "responds in crucial and defining ways to the nineteenth century's new media and the ideas they encouraged about information, communication and language."[13] Menke is not just interested in the figurations of the post office in literature. His argument is that new media technologies, chiefly those concerned with the transformation of writing such as the postal service and the telegraph, established the cultural and economic conditions for the genre of nineteenth century realism to flourish. As he writes in the introduction, "The point of this study is not only to discuss fiction that weaves plots around particular Victorian media technologies.... Rather I have sought to delineate the deep ways in which new technologies... register in literature's ways of imagining and representing the real."[14]

Realism weathered something of an assault in the twentieth century from critical perspectives informed by poststructuralism; what's interesting to me is how the postal service has also been used as evidence for the rise of a decidedly "unrealistic" style of writing. In the mid-twentieth century, realism appeared to lose its luster as a creative and critical force with which to make sense of the world. Across literary, historical, technological, and sociological fields it was attacked for the way it hid its artifice and presented itself as a natural "window on the world." Denying its own construction, realism could advance as natural particular perspectives about economic or social structure. Bound up in these critical movements was a gnawing sense that "representation" was no longer adequate as an analytical tool for understanding media cultures and visual regimes of meaning.

Artistic expression during this time echoed such suspicions, and the postal service reappeared in fictional and critical works as a way to draw attention to the deceptive modes of representation, to call out realism. Here we could include *The Crying of Lot 49* by Thomas Pynchon (1965).

In the novel Pynchon weaves real postal facts, such as the existence of the Thurn-und-Taxis Post, a nineteenth-century German mail company, with a fictional postal service called Trystero and its strange network of mailboxes disguised as rubbish bins.[15] And in a list of twentieth-century literary reactions to the postal service we can include Jacques Lacan's "Seminar on 'The Purloined Letter'" (a commentary on Edgar Allan Poe's short story), Roland Barthes's *A Lovers Discourse*, and Jacques Derrida's *The Postcard.* The letter is used figuratively and materially in these titles to stand for the difficulties encountered when imagining communication to be a straightforward enterprise. In *Special Delivery: Epistolary Modes in Modern Fiction* Linda Kauffman explores how the post office offers a compelling imaginative lens through which to see the interplay of radical "oppositional" readings of realism by examining a range of twentieth-century epistolary texts.[16]

Mail Art

That the postal service has been used to illustrate such a diverse range of creative and critical practice throughout the centuries is testament to its crucial role in everyday life. If literature explores the imaginative possibilities of letters through the epistolary form, then it is in "mail art" where the material structures of the post office itself are put to the test. Chuck Welch, one of mail art's central artists, defines the genre as

> an enormous, pluralistic, global communication phenomenon. Senders and recipients from around the world daily exchange expressive, provocative mail by recycling the contents and altering the surfaces of mailing tubes, envelopes and parcels. In time, these original, collaged surfaces resemble layered palimpsests of artist postage stamps, rubber-stamped images, cryptic messages, and slogans.[17]

Hunting for the definitive origins of mail art is tricky. In a sense, the postal service across its entire history has provoked aesthetic responses like the ones I've just discussed in relation to fiction. So the trajectory of mail art could include the invention of the pictorial envelope during the 1840s and the introduction of the picture postcard in the late nineteenth century. Mail art, sometimes known as "correspondence art," has been identified with the avant-garde movement; in its historical path also sits Marcel Duchamp, whose postcards were used as "readymades, distributing his ideas through the postal service."[18] Similarly, the date paintings and postcards sent through the mail by the Japanese conceptual artist On Kawara could also be considered mail art.[19] But it is the Fluxus movement and, more specifically, Ray Johnson and the emergence of the New York Correspondence School in 1962, where mail art begins to develop as a discrete art practice. In 1970, Johnson curated a mail art exhibition at the Whitney Museum of Art based on works sent in response to a call for entries—"letters, postcards, drawings and objects"—promising that everything received by the Whitney would be exhibited.[20] The show attracted mixed reviews, in part because of assumptions about the critical tensions that were revealed between exhibiting in a fixed gallery space and utilizing the flow of a postal network. Writing in *Artforum*, for example, one critic remarked how there was an inherent problem with exhibiting "a living thing in flight" while endeavoring "to pin it down and make a museum display of it."[21]

To exhibit mail art in an institutional space often seems to run counter to its egalitarian principles and open ethos. Requiring only a post box for participation, mail art functions as a significant form of public art. So, as John Held has argued, mail art often occupies a profoundly problematic relation to the commercial art sector. For Held, mail art has historically "distanced itself from the machinations of the art world, embracing an open structure and allowing access to anyone with an interest in the field. Even the term 'artist' was frowned upon, deemed elitist." Instead terms like "cultural worker" or "networker" have often been adopted.[22] Yet, the dialectic between art and commerce has always produced contradictory moments, as John Jacob notes: "Unfortunately mail art has a problem. It wants to be considered important; one of the High Arts. At the same time, however, it wants to be democratic available to anyone who has access to a postal service."[23] To a large degree, this open, distributed structure can threaten the longevity and historical record of mail art. By its very nature, mail art is ephemeral; the stamps, paper, envelopes, and postmarks of daily exchange pose a particular challenge for cultural heritage. And again, there is the potential to reduce public access by fixing the work in place, a fact Held eloquently captures in the title of his paper "A Living Thing in Flight: Contributions and Liabilities of Collecting and Preparing Contemporary Avant-Garde Materials for a National Archive." In this piece, Held struggles with his own conscience as both mail artist and librarian while he prepares to donate his significant collection of mail art and zine culture collected during the period from 1976 to 1995 to the Archives of American Art. Plastered on a postcard ready for submission to the collection, he notes, are the words of artist Carol Schneck: "Mail art is not museum art: don't file it."[24]

Many of the original mail artists of the 1970s and 1980s continue to make work, and in 2010 the British Columbia artist Ed Varney curated a substantial international exhibition titled "Mail Art Olympix." Exhibitors were invited to submit pieces by post that met the criteria in one of three aesthetic categories: "selfportrait," "artiststamps," and "manifesto." The resulting exhibitions displayed work by 340 artists from 41 different countries.[25] In turn, Varney's work itself is part of a recent large-scale digital heritage project, the Lomholt Mail Art Archive. Named after the video and mail artist Niels Lomholt, the project involved an exhibition called "Keep Art Flat! Mail Art and the Political 1970s," which was held in 2016 at Kunsthal Charlottenborg in Denmark. On show were more than 250 pieces of mail art Lomholt had collected from 1975 to 1985; these have subsequently been built into a comprehensive archival resource.[26]

Over the intervening decades mail art practice and archival efforts have continued to animate the imaginations of artists worldwide. Founded in 2005 by Paulo Magalhães, the Postcrossing Project launched with a unique goal to allow its members "to send postcards and receive postcards back from random people around the world. That's real postcards, not electronic!" Connected through a website and a number of social media forums, Postcrossing works on an exchange basis: interested participants request a postal address and then mail a postcard to it. Once that postcard is received and registered online, the initial poster may then receive a postcard from another user. As the website instructions explain: "You are now in line for the next person that requests to send a postcard. Where the postcard comes from is a surprise!"[27] Since its inception Postcrossing has amassed some impressive figures. Spanning 208 countries with nearly 700,000 members, the project has seen over 40 million postcards exchanged with around 400,000 postcards traveling at any one time amounting to 305 postcards sent every hour. Postcrossing may eschew electronic postcards, but email is the primary way by which physical addresses are made known to correspondents.

A similar initiative called Mailart 365 was launched in 2010 and describes itself as "a year-long project" where participants aim to "make and send a piece of art in the post every day for a year." For Mailart 365, the intimacy of the postal service is key: "in this age of digital communications, these personal touches and connections are getting harder and harder to find, so mailart looks to keep this tradition alive." Not everyone who contributes necessarily aims for of a full year of participation. Those who do, however, are called Mailart 365 Completers, and their work is displayed through the website and awarded with a special stamp.[28] One of the most prolific artists figured in Mailart 365 is Andy Salterego, who returns undelivered mail to senders, helpfully illustrating the envelope. On some occasions he sends elaborately decorated envelopes back to phone companies and banks, one being a beautifully drawn VW campervan complete with surfboards. Salterego laments that he "never had a comment back from them, but it's not going to put [him] off trying."[29]

It is quite difficult to do justice to the rich terrain of contemporary mail art projects. Jostling for attention are intriguing works found in projects such as *The World's Smallest Post Service, External Heart Drive, The Dear You Art Project, Postcard from the Past, Edwardian Postcards,* and *Letters of Note.*[30] One project that deserves special mention has captured the imagination and creativity of people around the world. Launched in November 2004, Frank Warren's *PostSecret* is described as "an ongoing community art project where people mail in their secrets anonymously on one side of a postcard."[31] It began on a small scale when Warren himself distributed 3,000 blank postcards at subway stations and art galleries, and inserted them into library books as well. People were invited to inscribe the postcard with a secret, perhaps a "regret, fear, betrayal, desire, confession, or childhood humiliation," and then mail the card to his address in Maryland. Participants were instructed to "reveal *anything*—as long as it is true and you have never shared it with anyone before." Not quite knowing what to expect, Warren discovered that "slowly secrets began to find their way to my mailbox." From around the world people were sharing their confessions often illustrated with arresting and poignant images. Since its inception, *PostSecret* has generated more than half a million messages displayed across art exhibitions, four printed and edited book collections, lecture tours and TED Talks by Warren himself, social media accounts, and two iPhone applications.[32]

PostSecret, Mailart 365, and Postcrossing all have a strong social media presence where contributors display their mail-art pieces, seek feedback, and discuss future projects.[33] Indeed, social media is breathing new life into mail art. On the one hand, as we are often told, the internet and email have killed off the post office with letter traffic on the decline. But on the other hand, across Twitter, Instagram, and Tumblr, postal projects are highly visible. Hashtags such as #mailart, #vintagepost, and #snailmail reveal images of beautifully crafted pictorial envelopes, vintage stationary, and pictures of traditional postboxes together with discussions about classes and workshops to teach the "art of snailmail."[34]

ASCII Art and Email Signatures

Although I have mapped precursors to the art of email in discrete sections, the development of many of these movements and projects overlap. While mail art grew in popularity from the 1960s to the 1980s as a form of creative expression using found or marginal objects, typographic experimentation was also taking hold. Many of the mail artists of this time were inspired by the avant-garde, mainly Futurism and Dada, and the subsequent flowering of visual poetry. It's possible to see the genealogy of visual or patterned word play stretching back before the twentieth century, but a key figure here is Guillaume Apollinaire, whose 1918 book *Calligrammes* helps to anchor our interpretations of ASCII art. (The abbreviation ASCII, as I mentioned in chapter 1, stands for American Standard Code for Information Interchange, a mainstay in electronic communication.) As Tom McCormack explains, in this work, Apollinaire aims to "create pictograms in which the text echoes the image." For McCormack, a key link to ASCII art is provided by how these early typographical works can become the "vernacular of the modern city," the ephemera and technologies of everyday life used to create meaning. This is why, argues McCormack, street art or graffiti helps to frame understandings of ASCII art and its development.[35]

Tracing lines from Apollinaire to ASCII art, a number of commentators note the significance of "typewriter art" through what Alan Riddell calls the "technological commonplace" of the office typewriter.[36] Like mail art, typewriter art was sometimes thought to be so mundane and niche as to be undeserving of serious or commercial aesthetic consideration. Yet the last few years have seen two substantial book collections of typewriter art published along with a renewed cultural interest in the typewriter more generally. Again, in ways similar to mail art, such creative focus is often inflected by technological nostalgia: typewriters crowd the popular consciousness from the screens of *Mad Men*–inspired iconography to retro-inflected music videos. But as Barrie Tullett has shown, for typewriter artists themselves "nostalgia" can often present a barrier to creative expression. "I'm not an 'analogue nostalgic,'" explains artist Dirk Krecker as he distances himself from the recent fetishization of technology consumption.[37]

Tullett conducted interviews for his 2014 book *Typewriter Art* with graphic artists, filmmakers, activists, and poets to explore how typewriter art has often remained a "low profile" within the arts even though recent works have enjoyed a resurgence of interest.

Similarly, in the 2015 book *The Art of Typewriting* by Marvin Sackner and Ruth Sackner, the authors explain the attraction of their object of study: the "beloved typewriter—its utilitarian beauty, the pleasing percussive action of striking its keys, the singularity of the impressed page—is enjoying a genuine renaissance across the creative industries."[38] A beautifully produced work of lush design and imagery, each cover apparently carries a unique picture so no two books are the same; the publication itself is drawn from the vast artistic archive of the Sackners. Established in 1979 at their home in Miami, the archive boasts the world's largest private collection of visual poetry but also encompasses other movements such as Surrealism, Bauhaus, and mail art. Comprising more than 60,000 objects, the archive includes works by Apollinaire, Allen Ginsberg, Roy Lichtenstein, Tom Phillips, Katharina Eckhart, and Gertrude Stein. Typewriter art is, in fact, just one aspect of their vast archive, which houses "experimental calligraphy, correspondence art, stamp art, sound poetry, performance poetry, micrography, … 'zines,' and graphic design."[39]

Having set the scene by considering some of its prototypes, I want to look at another marginalized or fringe creative expression, that of ASCII art and email signatures. One of the best scholarly analyses of ASCII art comes from Brenda Danet, who has attempted to produce an exhaustive list of its genres and styles. She defines ASCII art as "visual images created from typographic symbols on the computer keyboard."[40] How ASCII art is composed involves "a geometric medium," because every "image is a grid, constructed line by line, with each 'slot' or space in a line either filled by a symbol or left empty." Drilling down further, Danet uses art history categories to make a distinction between representational images—"drawings or pictures of objects in the physical world"—and abstract patterns. By far the most popular images in her study were representational, where green computer screens were adorned with famous figures like the Mona Lisa, or with machines, buildings ("castles are a favorite"), animals, and cartoons and fantasy creatures (like mermaids and witches).[41] Danet sees the zenith of ASCII art occurring between 1987 and 1995 as distributed across bulletin boards and email. She reminds us that it gained special prominence through the 1998 film

You've Got Mail!, when Warner Brothers launched an ASCII gallery featuring images of its leads, Tom Hanks and Meg Ryan, and invited the public to contribute their own designs. Along with the ASCII gallery the website provided an email "inbox" through which one could read the "actual" emails exchanged between the characters, Joe Fox and Kathleen Kelly. (At the time of writing, the ASCII gallery remained accessible on the site and was well worth consulting for its historic value as a rudimentary prototype for multiplatform film promotion, but it has unfortunately been removed.) The original site offered screensavers, virtual tours of filming sites, interviews with its director, Nora Ephron, and the option of sending to a friend a "webcard" illustrated with a film still. Moviegoers were also invited to submit email encounters of their own: "Have you got a story about a relationship that developed out of an email correspondence? Perhaps with a romantic twist? We took submissions (via email, of course) for a collection of stories about people who have met online." These stories, some complete with bridal snaps, were also archived on the site.[42]

Danet estimates that during its peak in the late 1990s, tens of millions of people were incorporating ASCII art into their email signatures or sending an occasional ASCII attachment via email. Part of the reason for its widespread use was the arrival of the Web, which made downloading "how to" guides easy, thus increasing ASCII's dissemination as a vernacular communication system. While ASCII art was highly visible at the time, Danet notes there were few substantial collectors or archivists maintaining stable historical records of the genre.[43] Since Danet's study, work has occurred in archiving ASCII art. Several, fairly stable, sites exist such as the "Signature Museum" and "ASCII World," which house historical examples of ASCII signatures and art more generally.[44] One of the most prolific artists is Joan Stark—Danet calls her the "Queen of ASCII Art"—whose work is still readily available in many of these archives (see figure 7.1). In addition, contemporary scholarly interest in meme culture and GIFS means that ASCII art has resurfaced as a significant historical stepping stone for these everyday public communication formats.[45]

Sites and Practices of Email Art

Having sketched some broad aesthetic frames, I now want to give a flavor of what I mean by email art. If there's one feature that links all the various

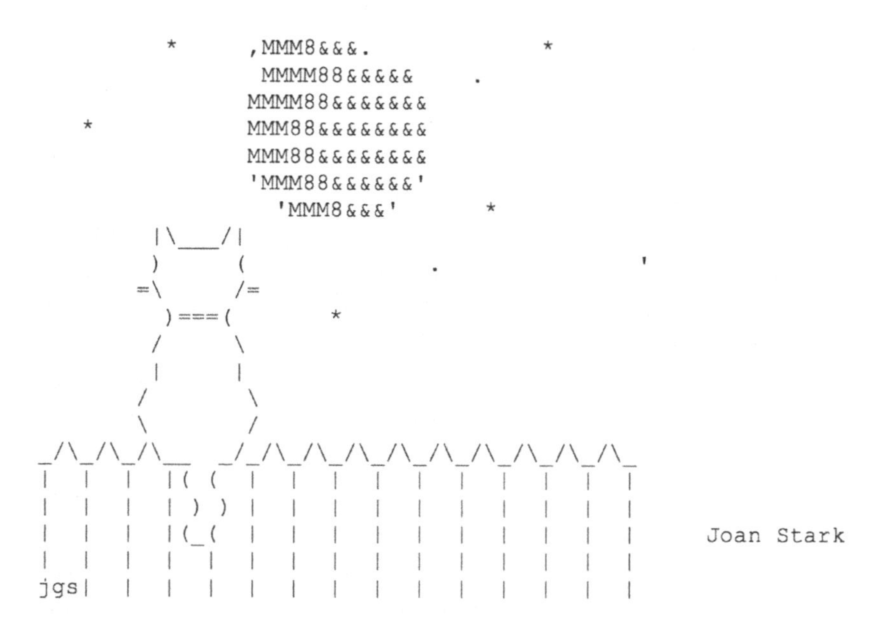

Figure 7.1
Cat Series by Joan G. Stark. Reproduced with permission from the artist, © 1996.

expressive forms I've discussed it is grassroots creativity and exclusion from traditional aesthetic institutions. It's true that many see eighteenth-century literature and epistolary fiction—*Julie, ou la nouvelle Heloise* by Jean-Jacques Rousseau, for example, or *Les Liaisons dangereuses* by Pierre Choderlos de Laclos—as highbrow. But the epistolary genre itself—stories told through letters—has in fact been relegated to the sidelines throughout history, often dismissed as a feminized or domestic genre.[46]

Another strong theme in the trajectories of artistic expression I trace here concerns everyday domestic intimacy. When communication technologies become deeply ingrained in our daily lives, creative practice can highlight the quotidian register, making us at once recognize its familiarity but feel oddly distanced by the sudden intense focus. A number of the email works I discuss below are animated by this double logic of domestic intimacy. In moments of defamiliarization, we can recognize the everyday look of screens, keyboards, and email software while seeing it anew in the spectacular encounter of a gallery space.

For each of the works explored below I spoke with the artists and content creators to learn their motivations for selecting email as their inspiration.

Carl Steadman: *Two Solitudes, an e-mail romance* (1994)

As I've argued throughout this book, reading other people's emails is a powerful trope across the internet. Whether that's through WikiLeaks-style distribution architecture or the quick peek over your colleague's shoulder at their screen, discovering what people talk about to each other in their personal moments can be a fascinating if clandestine activity. And for centuries it has figured as a persistent theme in epistolary fiction. The story of how the letters we are reading came to be published, how they fell into the "editor's" hands, is as much part of the genre as the content of the letters themselves.

In 1994, as the use of email became more commonplace, the writer Carl Steadman launched a project called *Two Solitudes*, which he describes as an e-mail romance. Interested participants could subscribe to receive in their own mailboxes a series of emails purported to be the exchanges between two romantic partners, Lane Coutell and Dana Silverman. The protocol was similar to subscribing to a mailing list or newsletter: to sign up the participant sends a "subscribe" email, and upon receipt a central email address would then be responsible for automatically distributing the mail. But the compelling aspect of this project was seeing the protagonists' names appear in the "to" and "from" fields. We, the eavesdroppers, felt as though we were reading someone else's mail; alongside our work email and spam, were the emails of Lane and Dana as their romance blossomed and then hit rocky weather. The welcome email from Steadman explained it like this:

> Thank you for subscribing to Two Solitudes, an e-mail romance.
>
> Over the next several weeks, you'll receive carbon copies of messages exchanged between two persons familiar with each other, as they send them. If their romance does come to an end, it is only because all things must come to their ends.[47]

The email correspondence between Lane and Dana spanned about two months, from September 24 to October 29, 1994, to be precise. But another engaging aspect of the work is that participants would receive the emails with the real current date of receipt displayed. A reviewer of Steadman's piece published at the time wrote that the technical design "looks to anyone who is subscribing that they are actually reading someone else's mail. It's a remarkably intimate feeling, despite the fact that it takes place over a computer."[48] In their emails, Lane and Dana discuss work, going shopping, recipes, song lyrics (Pet Shop Boys are a favorite) and their relationship. Their conversational tone is familiar and routine; Dana tells of work as a temp, Lane's cat needs to go to the vet. The first exchange sets the scene for their separation:

> Date: Sat, 24 Sep 94 15:36:20 CDT
>
> From: Lane Coutell
> To: Dana Silverman
> Subject: hello...
>
> Dana-
>
> I am writing this to you, so that when you first access your account, you will have mail waiting for you. I hope the new setup works out for you.
>
> You only left today, Dana, and I already miss you quite dearly. I hope things work out with your mother, and that you'll write me often. Three months seems like a long time—and will I even see you then?[49]

Despite their loving emails to each other, sharing their everyday activities, wishing they were together, things did not pan out well for the couple. In a nicely underplayed, deft gesture Steadman has Dana explain, seemingly out of the blue, a desire to be alone. "I need sanctuary." Sadly this proves to be the last email Dana sends. As Lane loses contact, the final email of the project that we receive reads:

> Date: Sat, 29 Oct 94 08:15:21 CDT
>
> From: MAILER-DAEMON@sobriquet.com
> To: Lane Coutell
> cc: Postmaster@sobriquet.com
> Subject: Undeliverable mail
>
> Your message was not delivered to the following recipients: dsilverman: User unknown[50]

Over the course of the project's life, which ran as an active subscription service for seven years, it attracted about 30,000 subscribers. Although this may seem modest by today's social media numbers, it does represent a significant level of interest. For Steadman it was "one of those things that had its own little corner of the internet." Part of the attraction was its adherence to representing the sociotechnical affordances of a real email encounter. Looking back, Steadman sees this as key to its success as he explained to me:

> I think what set *Two Solitudes* apart, and what compelled me to write it, was that it was episodic without being periodic. It was a correspondence between two people that would proceed at its own pace outside the reader's control. It might be a day

or two before a particular missive would receive its reply; as a reader, you could do nothing but wait, so that the whole story took over a month to read. It well captures the closeness and remoteness of personal email in the age of dialup.

Two Solitudes was subsequently published in the online fiction magazine *InterText* that ran from 1991 to 2004. And *InterText* has archived Steadman's work, so the emails can be read in their entirety.[51] Although obviously good for cultural posterity—and I will talk more about archival practice soon—this format belies the original site-specific intimacy generated by the daily arrival of emails. Steadman himself says it actually took some convincing for him to agree to its publication in this format. As he explained when I asked him about the archival situation, "I still don't see how it works in any form but its original. To me *Two Solitudes* derives almost all its charm and delight from the voyeuristic thrill of reading other people's mail."

Miranda July: *We Think Alone* (2013)

While Steadman's work is fiction, Miranda July's project deals with real emails. Again, the email delivery system itself was used as part of the artwork, with the final pieces only fully available as emails arriving in people's mailboxes. *We Think Alone* was commissioned by the Magasin III Museum for Contemporary Art in Stockholm as part of its 2013 exhibition "On the Tip of My Tongue," curated by Richard Julin and Tessa Praun. For this work July persuaded her friends in the culture industries—artists, celebrities, filmmakers, actors and writers—to participate by making available their private email correspondence. Contributors included the American actor Kirsten Dunst; the creator of the hit TV program *Girls*, Lena Dunham; the photographer Catherine Opie; NBA sports legend Kareem Abdul-Jabbar; and the Israeli filmmaker Etgar Keret. To access the emails, people signed up through July's site and were then sent a weekly email digest containing the participants' correspondence grouped around themes like "An Email about Money," or "An Email You Decided Not to Send." During the course of its 20-week performance the project attracted over 100,000 subscribers who could learn about, for example, how Lena Durham was considering buying a $24,000 couch.

Discussing the genesis for the project with me, July explained that it built on a previous work in which she had also tapped into the creative potential of email:

As part of the promotion for *The Future* (2011) I invited the public to sign up and have their "future" emailed to them twice a week. I wrote a few hundred divinations that were randomly sent out each week. It went on for years. Magasin III, in Stockholm, was aware of this project and asked if I might do something similar for an exhibition they were working on that extended beyond the gallery. I was wary of taking on something that involved writing (since I was in the midst of writing a novel) but thought that if I could use "ready made" emails then it might not be so hard (of course it was much harder in the end.)

Not surprisingly the project generated media interest because of its high-profile participants. In many ways the aim of trying to capture the routine or everyday nature of email does sit in tension with the necessary editing involved and the fame of those involved. I wanted to hear more about the celebrity aspect of the project and asked July to reflect on her aesthetic choices. As she explained:

I am always interested in reading my friends' emails to their mom, agent, husband, etc.—I like to see how they relate to the other people in their life. So the decision to involve celebrities was based on trying to re-create this friend feeling. I also wanted a diverse group—a physicist, an athlete—not all just writers and actors. And I wanted at least one person who would be more visual (Catherine Opie.) It was a bit like casting a movie—you're trying to choose the people who will tell a really interesting version of the story, as a group. The fact that so many people subscribed (because of the celebrities) also turned out to be important, it meant there could be a weekly conversation, as with a TV show. Kind of interesting for something so non-narrative, conceptual and uncommercial.

I was also interested in the decision she made not to archive the project. Media outlets had only published highlights rather than the full emails. What this means is, as the exhibition curators explain, there have been no formal archives of the emails since the project "only exists in your inbox."[52] I asked July whether she had to balance wanting to record the project with its conceptual, site-specific objectives:

Yes, exactly. If you make the emails into a book, suddenly it's not really in the medium of email any more. I liked the fact that the subscribers, all 100+ thousand of them, were the only owners of the work, forever. It's ephemeral, or not, depending on if you decide to keep them or delete them, and depending on how long the medium lasts. I still see people posting calls on social media for the complete set of emails.

Although July has sometimes been criticized for the celebrity feature of the project, with the emails seen as mere PR and too obviously edited, the cultural fields of visual art and fame already have a long-established mutually informing tradition stretching back at least to the 1960s: an aesthetic practice that has explicitly sought to explore and expose the complex interdependence between cultural visibility and art. This element of the dialogue is of less concern to me, however, than what the project says about assumptions of email communication. July's work came to us before the so-called Sony Hack in 2014. As I briefly noted in the previous chapter, this event concerned the unauthorized publication of internal Sony emails that made very public some of the highly controversial exchanges between the then Sony Pictures co-chair Amy Pascal and the producer Scott Rudin. Also revealed were personal emails involving high-profile actors such as Angelina Jolie and Leonardo DiCaprio. It is difficult to imagine a similar project to July's being launched to much fanfare in a posthack environment.

Daniel Smilkov, Deepak Jagdish, and César Hidalgo: *Immersion: a people-centric view of your email life* (2013)

Taking a different approach to the two works already considered, the creators of *Immersion* put the focus squarely on our own inboxes rather than letting us peer at the contents of others. Launched in July 2013 at the MIT Media Lab by Daniel Smilkov, Deepak Jagdish, and César Hidalgo, the *Immersion* project is a visualization engine that maps connections using Gmail metadata, specifically "the From, To, Cc, and Timestamp" instead of "the subject or the body content of any of your emails."[53] Participants submit their Gmail address and password so the program can scan their emails, up to 30,000 at any one time, to draw a network diagram of the communication patterns at work. As is the case with social networking visualization, *Immersion* shows the volume and frequency of communication by emphasizing those with whom participants engage regularly, and it charts the strong or weak ties existing between the other people in one's mailbox. After a map is produced, users are given the opportunity to delete the metadata from the *Immersion* servers and logout, or they may choose that the data remains accessible so they can further develop the picture of their social relationships. *Immersion* reminds us about the longevity of email and its significance for cultural heritage purposes. For the creators of the project, it is motivated by the recognition that

> email, one of the original forms of social media, is even older than the web and contains a detailed description of our personal and professional history. *Immersion* is an invitation to dive into the history of your email life in a platform that offers you the safety of knowing that you can always delete your data. Just like a

> cubist painting, *Immersion* presents users with a number of different perspectives of their email data.... It's about providing users with a number of different perspectives by leveraging on the fact that the web, and emails, are now an important part of our past.[54]

Since its launch *Immersion* has attracted nearly 1 million users, people keen to see what their Gmail box says of their digital lives. For Hidalgo, the strength and sustainability of email lies in its accessibility: "It is a protocol, not a platform," he argues. "You do not need to have an account with the same platform to communicate. I cannot send you a message on Facebook if you are not on Facebook, but I can send a message from my MIT email... without worrying about who is running that email server."

Immersion was launched amid the revelations of surveillance carried out by the US National Security Agency (NSA) and exposed by Edward Snowden in June 2013. Snowden had leaked to the *Guardian* evidence that US government programs were routinely tracking phone calls, internet use, instant messaging, and email communication of the American population as well as having monitored embassies worldwide. Key to defenses of such surveillance practices is the explanation that what is accessed is simply metadata, data about data, and the content of people's interactions are not being traced. When similar policy was introduced in Australia during 2015 under new metadata retention laws, government ministers and those in support of the policy argued these measures were justified because they would help combat organized crime and terrorism. This legislation requires telecommunications providers to retain for two years records of customers' online and phone activities, which would include time, date, location and duration of phone calls, email addresses, sizes of email file attachments, and the time, date, size, and recipients of emails.[55]

As many commentators have noted, metadata is far from neutral and, in some ways, offers much more detailed profiling capacities than do those monitoring systems that focus on content of phone calls, texts, or emails. Indeed, as the NSA's general counsel, Stewart Baker, has himself bluntly stated, "Metadata absolutely tells you everything about somebody's life. If you have enough metadata, you don't really need content."[56] For José van Dijck, metadata has become a form of currency whereby citizens and consumers trade their personal information—with governments for security and with commercial platforms for social relationships. Running beneath these transactions is a set of assumptions about the objectivity of data;

instead, as she puts it metadata "are value-laden piles of code that are multivalent and should be approached as multi-interpretable texts."[57]

In *Immersion*, the value of metadata is made concrete. Although he does not intend his projects to critique NSA and other global surveillance programs, César Hidalgo remarks that he has actually been working on information aggregation projects for many years in order to bring to life the power of data science. "It's like the world is catching up to what a fringe group of academics was aware of in 2004 and 2005," he comments. Back then people didn't worry about metadata, he says. "Nobody cared about us, and they all thought that working with mobile phone records or e-mails was sort of a curiosity or a stupidity.... We've come to a world where, now, it's completely the opposite. Everybody's chasing that."[58] Hovering between art and science, Hidalgo and his team are also responsible for another data visualization engine that takes the Hillary Clinton emails, discussed in the previous chapter, to produce what they call the "ClintonCircle." Using the tools developed from the *Immersion* initiative, the ClintonCircle is a massive data visualization interface that lets users plot Clinton's email traffic to show its volume, frequency, and direction. The database links to the WikiLeaks archive so that content of emails is viewable too.[59] As with Enron, one is struck yet again by all the ordinary people who are suddenly scooped up into such projects.

Brian Fuata: *All Titles: The Email and SMS Text Performances (2010–2016)*

Literally embodying the affective and creative possibilities of email is the work of Brian Fuata. The Sydney-based text and performance artist has been exploring what he calls "electronic correspondence as a theatrical stage" in a sustained effort over a number of years.[60] For these pieces Fuata imagines the address fields of email, the Cc and Bcc most importantly, as areas for listening and watching, a "scenographic space." Using the metaphor of the theater he shows how email creates publics with minor acts of theater that occur in our mailboxes. As with some of the other projects I've looked at in this chapter, Fuata's performances involve people signing up to receive his stagings, which are scheduled in semi-regular intervals. The name of the particular piece will appear in the subject line: *Email Performance Act Three*, for example. As he explains:

> Since 2010 I have been sending what I call "email performances," which are/become performance texts that I send directly "To someone/group" while also

> including a selection in the Bcc field; that I liken to the darkened space of the audience in a theatre. In my mind, the To field places the person/s in an odd place of public show; implicating whoever as a fellow performer as well as witness to their subject-objectification.... The category of the Cc field lies between a strange performance place of witnessing. I liken the scenographic performance space of Cc to the aesthetic works of much post-dramatic theatre and dance throughout the early 90s.

Fuata's work focuses on the ordinary encounters of everyday life, those tiny moments of interaction that stitch our lives together, which are then juxtaposed with theatrical spectacle. As I've discussed in relation to the photography of Lars Tunbjörk, we can see a double logic of intimacy operating within Fuata's media arts practice. On the one hand, there is the gravitas of an art show; Fuata's training in dance and theater is palpable as the emails we receive from him are often quite dramatic in their tone. Many of the emails begin with stage directions such as "curtains, rise, open." In *Email Performance Act Five: A Nervous Body*, the tone is portentous: "the beginning of the end; the beginning often the end; the begging for the end," he warns. And yet in other emails he is conversational and matter of fact, sending us pictures of cats. Maintaining an air of verisimilitude is important to Fuata. He tells me he never uses bulk-emailing services like Mailchimp but always sends from his own email address.

Email Performance Act Six: A Busy Body combines both senses of intimacy as extraordinary and ordinary. Bear in mind that I don't know Fuata (although in researching the book I talked with him about his art), so these highly personal images and prose make a significant impact. Here we encounter pictures of leaves falling in upstate New York where, we are told, he has gone at the invitation of his friend Dan. Perhaps that's a picture of Dan himself, standing in a very ordinary kitchen wearing a singlet. Echoing the epistolary tradition, where letter writers often draw attention to their correspondent's absence, Fuata captions the image by saying:

> a thanksgiving coffee with dan taulapapa mcmullin, who invited me into his home in hudson, new york where i am til sunday morning. dan is also here in the to field. i imagine him now, despite the fact he is next door reading his book in bed ~ and i will meet him in the morning for a mountain hike, and i will still imagine him as he is in the to field as separate to the dan who i laugh with love.

To me, this feels incredibly revealing. And I wonder if it is really any different to the thousands of personal disclosures we read every day across Facebook, Instagram, and Tumblr? One reviewer of Fuata's work thinks so.

As Gail Priest puts it, "These small sharings are... the very opposite of the wholesale banality of the sharing offered by Facebook."[61] Similarly intimate and slightly unnerving is the piece *A Performance for All the Men in My Address Book* where we can, indeed, see many email addresses cascading before us in a blue wall of hyperlinked names. "We" are in the Bcc field, again watching from the sidelines, as Fuata performs for his friends with poetry and song lyrics. As well as having been delivered to subscribers' mailboxes, the piece also exists in video form. Here a delightful domestic scene awaits while the cursor hovers over the familiar tabs in the Gmail mailbox. Fuata scrolls through the email introducing the various friends to each other, actual names that appear in the "to" field: "some of you know each other and/or me, others not at all." I was intrigued about the genesis of this piece and ask Fuata to explain further:

> I was in my studio and was suddenly struck with the idea of "the email shows" as a potential space for queer intervention, specifically relating to the male gaze. I riffed off a solo theater work I did in 2003, Fa'afafine, where I divided the audience into men and women. In the show I ushered the men onto a raised mini-stage of two blocks of rostered seating facing each other, in between was a little catwalk-like performance area where I performed texts about intimate sexual moments of male lovers etc. It created a small theater of them witnessing each other and me, and my implication of them as narrative prop/subject/object, on top of which was witnessed as a whole image by the women placed in circular, cabaret-style seating.... This was followed much later by a work where I separated audiences between white and People of Color instigated by a ... survey exhibition on the history of African American performance art somewhere in L.A. I mention this to illustrate that the emails have no set agenda, and that they're context-dependent. It's the context, which changes all the time, that guides my approach in construction.

Fuata has performed internationally and his work is strongly informed by mail art practices. In 2015 he exhibited at the Performa Biennale in New York. For these shows Fuata drew on the research he had conducted with the Ray Johnson estate and archives. As he put it to me: "The role of mail art changed my entire practice through articulating the idea correspondence as a method."

Conclusions

In this final chapter we have considered email as an aesthetic subject, although perhaps an unlikely one. In order to provide context for thinking about the art of email, I suggested a number of historical trajectories. The first is the established yet still sometimes controversial field of "digital media arts," where the chosen media is often the target or subject of the work of art itself. So, we have bioart, which uses biological materials as media to make broader points about contemporary ecological issues. This represents one of the most obvious facts about email as art: that it self reflexively explores the means of communication via that very communication itself. Thus in all of the email works I explored—*Two Solitudes, an e-mail romance* (1994); *We Think Alone* (2013); *Immersion: a people-centric view of your email life* (2013); and *All Titles: The Email and SMS Text Performances* (2016)—email was both form and content, the means by which the work was distributed and the subject matter as well.

The other historical location for email art is the postal system. As communication media, email and letters have been in dialogue since at least the late 1970s when the United States Postal Service (USPS) attempted to take over the business of email transmission. And as we saw in chapters 1 and 2, at various key points in the history of email, the post office has played a material part in defining the remit of correspondence carriers and the manner by which traffic should flow. Figuratively too, postal iconography infuses our screens as we interact with images of mailboxes, envelopes, and stamps. Indeed, one of the earliest email clients, Eudora, is named after the twentieth-century writer Eudora Welty and her short story "Why I Live at the P.O."[62] Such histories join us in the present as the USPS announced the "Informed Delivery" service in 2017, which scans the outside of your envelopes and sends you an email with an image of the contents.[63]

To pursue some of these connections further I explored the creative uses to which email has been put. Looking to its precursors I began with epistolary fiction of the eighteenth century to argue that technical infrastructures such as the post office have for centuries provided imaginative frames for thinking through what is at stake in our communication systems, how their architecture functions, and how content moves. Writing about the post office, marveling at its advances or grumbling at its delays, characters in letter fiction can be seen as nodes within a rapidly expanding information network of the era. So again, we can see self-reflexivity at play. It is no coincidence that the email novel emerged in the early 1990s just as email was becoming commonplace in workplaces and homes. And whatever you may say about the cinematic value of *You've Got Mail!* (1998), the film remains a significant instance of the dialogue between emerging and existing media.

Alongside email novels, mail art is an important site to consider when thinking about the lineage of email aesthetics. Perhaps one of the key lessons is how mail art opens up the possibilities for using public systems of communication to make meaning. Here we can see a mode of creative practice very much alive today in the various projects, both digital and analogue, that draw on the postal system as a muse, projects that include *PostSecret* and *Letters of Note*. In mail art, we can also find sparks of inspiration for ASCII art and the elaborate email signatures of the 1990s. But perhaps one of the most salient parts of the history for my purpose in this book is the way that mail art evokes the domestic and everyday, and in so doing creates moments of intimacy. With source material often ready to hand—paper, envelopes, post boxes—mail art draws attention to the commonplace or the routine of daily life. What I called the "double logic of domestic intimacy" is also at play in the email artists I discuss. This term tries to capture how works of art about the everyday often function through defamiliarization; we recognize the everyday media of, for example, the email header with its subject line or address, but see it afresh as it arrives in our mailboxes via subscription to the work, or when it is displayed in a gallery space.

Conceptually and materially, the archive underpins many of the art projects discussed. Throughout the chapter we have seen artists expressing different attitudes to the question of archival policy and practice. Often the very ethos of mail art, its radical insistence on open access and its eschewal of commercial exhibition paradigms, means preservation represents a particular conundrum for both collectors and artists. Some like Carl Steadman and his *Two Solitudes* piece bowed to pressure, agreeing to publish and archive his work. Others like Miranda July in *With Think Alone* resisted such overtures and insisted on the specificity of the site encounter. While email signature art might not have been so visible recently, these expressions undeniably paved the way for contemporary meme culture, which looks at the vernacular use of imagery. Luckily for scholars of such practices, there are in existence some fairly stable archives of these art works in the Signature Museum and ASCII World. Archives can help to tell stories of history through their documentation of particular moments of political or cultural significance, the letters of famous people being an enduring example. Email archives are perhaps still in their infancy but represent an important site for continuing aesthetic and cultural attention.

Conclusion

This book could have looked much different. One of the challenges in writing about a ubiquitous, everyday communication form is finding stories to tell that are novel enough to compel attention yet sufficiently familiar as to spark recognition. Email functions almost like a shared language; people seem instinctively to know what it is and, of course, what's wrong with it. At social gatherings and at work when chatting about this book, friends and colleagues often quizzed me on just what I would be covering. *Will you talk about that threading thing when you think someone hasn't answered? What about those people who send walls of text in their emails? Two lines max! And don't forget terms of use: does anyone really read them?* In response I would sometimes rush home and expand a particular point or chapter; other times I would shrug and say, "I can't include that, people know what email is without giving every last instance."

In a sense, this takes me to the heart of *Email and the Everyday*. Yes, we all know how email functions within our own sphere of normal work and domestic media engagement. Equally, in research many individual studies have investigated a vast array of email practices, beliefs, styles, and purposes. But how do all the parts fit together, how is the terrain constituted as an interconnected, media specific field of practice and consumption? Here then is what I have attempted to provide to scholarship: a sustained, critically and empirically informed map of contemporary email use in everyday life. As I have shown, email underpins so many of our daily interactions—commercial, political, personal, and transactional—that it can become invisible. In the remaining pages I want to revisit some of the key structuring themes of the book in order to summarize what has been argued about email use in contemporary and historical life.

Stories of the Everyday

When I began to notice that "stories" kept popping up as a theme in the research I had conducted, I assumed it would function simply as a heuristic, a method to organize the material coherently. In particular I was finding that explaining much of the historical, social, and technical detail—the minutiae of email protocol development or the reporting mechanisms of email marketing analytics—was obscuring the overall sense of email's significance. That had been, after all, the motivation for the book in the first place. I worried the book would submerge readers beneath a sea of detail (although I always find detail more compelling than grand theorizing) and lose them in that process. I remember quite clearly saying (out loud, I admit), "I want this to be like a podcast." By that I meant I was searching for a mechanism to make comprehensible or "relatable" all the disparate strands of email consumption and production. But as I began to reflect on the way to tell the story *about* email, I realized the stories were also being told *by* email.

During the course of this book we have encountered emails in many different situations. There have been emails written to share personal news with friends and family, those sent to arrange events, emails complaining of colleagues, or emails that offer support to fellow illness sufferers. Some emails are the work of careful crafting and drafting, reading several times before sending, but in others a mere keystroke will guarantee the contents are suddenly released unexpectedly to a large group of people. In most cases, you will know the people you exchange emails with. Even if you haven't ever met in person, email often creates feelings of familiarity as it joins together groups with similar institutional, cultural, political, or personal interests. But sometimes you may find yourself reading emails not addressed to you or, more broadly, accessing widely circulated databases of email subject material. In order to categorize such a diverse range of situations, environments, affordances, and applications I used the key terms "disclosure," "trust," and "digital labor."

Taking a step back, however, I found that in all these instances what distinguishes email from many other digital media forms is its capacity for storytelling. A salient example of this feature, as we saw in chapter 7, is the email novel. Building on the long narrative tradition of stories told in letters, the email novel, although perhaps marginal in literary criticism circles, has proved a productive narrative choice for fiction authors. And yet the narrative impulse of email is not limited to fiction. Across all those "real world" scenarios quickly sketched above are stories being written, read, shared, and published. More specifically, as we saw in chapter 5, within email discussion groups, narrative is a significant strategy by which many support groups are able to cohere. For those managing the health-based communities I explored, telling stories to each other about their situations is a crucial aspect of their communication practices. Not only does narrative figure in the content of emails shared between members of these groups, but I also discovered when talking to moderators that there are intricate and compelling stories about how these emails get posted on groups and lists in the first place. Asking moderators to reflect on how they came to assume their roles revealed quite complex stories of governance and labor.

The counterargument could be made that email is not unique in this aspect of digital media: forums such as Reddit and Whirlpool, or platforms such as Gamespot, Facebook Groups, Instagram, and Snapchat, all present opportunities for narrative expression with the two latter platforms badging themselves explicitly by offering "stories" as a feature. But none of these sites rival email for sheer user numbers, longevity, or nonproprietary ease of access. (As an aside, and as noted generally in the introduction, email is the prerequisite to sign up for every one of these platforms just mentioned.) In other words, one could make the argument that email represents the most widely spread and historically persistent expression of digital media storytelling. Perhaps this dimension of email has remained hidden when newer forms of narrative, such as podcasts or games, seem much more obviously to carry the "digital storytelling" mantle.

Yet as I have tried to show throughout this book, email as story operates in clear and compelling ways that are at work on three levels: first, the stories of email refer to what people say in their emails to each another; then there are the stories of email circulation, publication, or archival, with Enron and the Clinton emails being two examples; and finally there are the stories told *about* email in our public conversations, where they might refer to the death of email or its part in work/life balance, email overload, or its insecure protocols and the massive hacks and breaches.

As a final point to this section I would add that email is the dominant way that the public can share their own personal stories on well-known radio programs or audio documentaries such as *This American Life* or the Australian *Changing Tracks* radio segment (aired by the national broadcaster

ABC, the Australian Broadcasting Commission). *Changing Tracks* asks audience members to email a story of "a song that was playing when your life changed track."[1] The website for *This American Life* hosts an archive of emailed story pitches that eventually made it to air; they are displayed in email format, with date and subject lines of the actual contributors, and a link to show numbers.[2]

Now, I don't want to overplay this point or make too much of a transmission method that could easily change; clearly, media broadcasters also often invite participation through their Facebook groups, Twitter pages, or SMS phone lines. But what I am trying to get at here is that storytelling is one of the ways in which email underpins our daily lives. What I have been trying to uncover is the means by which email becomes part of the everyday fabric, or is "saturated" (to use the terms of media theory). And email's capacity for stories seems key to this process. Crucial here too is that the storytelling of email is subtle, it doesn't happen in every single email we send. But its ability to offer users a way to communicate narratively is one of the ways email becomes part of the everyday. It is also perhaps a reason for the relative longevity of email; that it can be used to cover transactional or merely informational situations and, as I've just explored, it also offers an environment for intimacy and sharing.

To close this section, I want to reflect briefly on lessons learned about "the everyday." As I discussed in the introduction, the everyday has been used to analyze the routine nature of our encounters with a broad range of consumption practices including media. What Ben Highmore calls "absent-minded media" is a way to show that questions once posed of communication technologies through processes of "representation" have now shifted to those of "attention."[3] And, as he argues, that attention itself doesn't operate homogenously. Certain media forms invite more focus at particular moments in history while others become so ubiquitous, so woven into the fabric of everyday life that we do not need to pay any attention, for they immerse us. In fact, as he points out, in times of intense media saturation we need *more* effort to avoid a particular media form. Clearly email is here a fitting candidate for such a description. Not checking emails, taking digital detoxes, these are works of conscious effort. But I also suggest that this attention functions at the meta critical level and the everyday worlds of media scholars. Email represents an odd media case, not quite niche or forgotten enough for the critical tastes of media archaeology to revive, but not new enough to attract the attention of most digital media scholars. Yet, as I hope to have shown, its ubiquity and global reach warrant continued analysis.

Histories

Observations about the attention of media scholars invite us more generally to look at the role of history in the exploration of email. I devoted two chapters to the history of email in an attempt to capture its work at two levels: first, the small and incremental changes to its protocols, including the format of email headers and what fields should be included; and second, observations about cultural or economic currents that have also affected its development, like the connections between the USPS and email over many years.

Underpinning many of these technical advances is the Request for Comments (RFC) series, the principle vehicle by which internet protocols are established. The RFC series was begun in 1969 and is still in use today by engineers and software developers, representing a vast publicly accessible database of standards literature. Here in these publications we find the early definitions of what would become the email attachment, or we can see burgeoning principles for combatting spam. During this period the initial steps were made toward what we recognize as email today. In 1971 Ray Tomlinson created the first network mail program and also selected the @ sign to distinguish elements of email address syntax. But these RFC texts are not "simply" technical discussions; instead the archive demonstrates how technological history and development is always a process of dialogue, negotiation, and conflict, often the result of differential access to resources and funding. In the early days of the internet those with less institutional support sometimes felt they were outside the decision-making processes. The "standards war" of the early 1990s, which waged as the key internet protocols were rolled out, is instructive for how it contests a smooth linear development of historical invention where each subsequent application simply builds effortlessly on what went before. Rather, as Andrew Russell has shown, at play were quite vigorous political, cultural, and personal battles about who could control and design the protocols for these new information systems.[4]

Since this book is about stories, I am of course very interested in the way historical narratives are themselves written. In 2012 I began to see curious historical accounts appearing in news items and on mailing lists about the "real inventor of email." Well acquainted with the critiques from science and technology studies about how communication history can often be presented as an inevitable outcome, the march of progress shored up by a technological determinist view, my interest was piqued. Moreover, as social constructivist historiography has taught us, what's often wrong with these narratives are the omissions, because what can be left out are stories from marginalized social groups whose contributions to technological history remain concealed and ignored.[5] So it seemed the perfect setting in which to consider the claims from V.A. Shiva Ayyadurai that it was he, not Tomlinson, who had invented email in 1978. As I discussed in chapter 2, this is the narrative that has been adopted by Ayyadurai himself as he advanced the idea that he has been deliberately omitted from the history books due to a carefully orchestrated conspiracy led by commercial interests and industry insiders. One of the problems with Ayyadurai's stance, let alone his actual claims, is that it leaves very little room for contributions made by the many others who worked in the computer science labs of the time. He may very well say he is ignored by the broad internet and computer science communities, but he rarely seems to accept their work either.

In email development, the slow evolutionary steps collaboratively made or the failed experiments and missteps taken fly in the face of histories written in the (often hyperbolic) language of absolutes starring "inventors," "pioneers," and "heroes." As Susan Douglas has shown, the role of journalism has been crucial in shaping such simplistic narratives of technological change that often privileges the revolutionary image of "firsts" over the messy reality of uneven historical development. To understand how certain devices are historically consumed and perceived, we need to look at their symbolic framings and recognize the press as a significant actor in the social construction of technology. More specifically, Douglas argues, it is ideology at play, which as she explains "is the prevailing and evolving common sense about a technology, as consolidated through media frameworks." Douglas encourages scholars "to examine how media coverage of technologies, often deeply contradictory and incoherent,...can shape their dissemination and uses."[6] Such recognition is why it seemed important to locate Ayyadurai's claims within the wider sociopolitical landscape of celebrity, media, and law.

But this presented a complicated discursive terrain to navigate populated by disparate social groups, industry players, and institutions. The risk in trying to weave together so many strands was the loss of detail; that the story now told would obscure many of the historical, material facts about the development of email. For some, perhaps even those from the science and technology studies' communities whose opinions I was seeking to convey, these various narrative threads—computer history, celebrity legal actions, technology journalism, Silicon Valley–backed law suits, and tabloid media's invasion of privacy—should all be kept distinct; each has its role to play within a highly diversified, distributed media landscape, but muddies the waters when viewed together in the proper history of email. Clearly, I felt slightly differently and hope to have shown that these historical, political, and cultural forces are not possible to separate cleanly.

Markets

One of the most surprising and increasingly frustrating discoveries I made has concerned the dearth of independent survey data about worldwide email use. By far the most cited source of email statistics is the Radicati Group, a market research company based in Palo Alto, California. Drill down on pretty much any news articles or jokey blog about information overload, or the etiquette involved in "ccing" your manager, and you'll find that the evidence for email usage derives from the Radicati Group.

Although "objective" or "independent" may be relative—how you ask a certain survey question obviously influences your sample, and recruiting respondents via landline might skew the age demographics—a commercial study will differ from research conducted at a university. Here I am referring to a large-scale, overall snapshot. How often do people check email? What is the most popular email client? How many emails are sent each day? And inevitably everyone wants to know the contrasts between email and social media use. Since email marketing is still a major purpose of email, it's not surprising that figures about its saliency emerge from companies in the business of email analytics. But as I explored in chapter 3 on the email industry, quite often these studies are motivated by a desire to prove either the longevity of email or how social media has supplanted it as a method to reach audiences. Of course this is not to suggest these studies have no value; indeed, again as I've shown, they can provide their own story about data collection and analysis. But the fact remains that it is exceedingly difficult to glean independent, broad-picture details about how people access, write, receive, and exchange email across different devices and clients.

Moreover, the use of methods such as cookies by commercial studies to track opens may seem a reliable or objective measure of email use, they don't tell you much about how people feel about their email software. Email brands might take pleasure in figures that show, for example, the rise of Gmail, but when people are asked to name their own software many remain fairly oblivious to such specificities. In my 2017 survey of email users in the United States, many people either incorrectly named the particular email program or used a variety of spellings and approximations: Google for Gmail, Outreach instead of Outlook, and others added descriptions such as, "work one," "company," or "university."

I want to make three interrelated observations about these responses. The first points in part to the ongoing need for communications research that tries to combine figures and feelings, an approach that is not caught up by the sway of numbers. The second points to the general issue of accessing and evaluating industry research. In many cases, the commercial sectors of communication and information providers will fiercely guard data about the market impact of their product or services. While such reluctance is understandable in terms of commercial confidentiality and, frankly, limiting the damage when their success is not all that they claim, it makes working with these groups quite challenging. As I noted in chapter 5, it was very difficult to find any statistics from Yahoo! Research about their user base—despite the fact I wanted to explore and, indeed, extol, the benefits of their own email discussion groups. As a number of media scholars have noted, collaboration with these communication and entertainment giants is often a tricky endeavor.[7] Yet we increasingly see these same organizations complain vociferously about their inability to meet community expectation. Perhaps relying solely on commercial research findings is hampering these companies' full grasp of the rich exchanges that play out on their platforms. It's just possible that university research might help. And the third observation—that people might not identify a brand with the precision that commercial studies would wish—also illustrates how the broad category of email operates beyond the discrimination of the market that is limited to individual applications. In other words, I'm talking about email as a system rather than instances of a brand, a point I now want to elaborate in terms of its relation to social media.

If I have spent most of the pages of this book outlining the pitfalls of email—how its constant insistence encroaches upon our leisure time, or that trust is misplaced in how it protects confidentiality—perhaps now is the time to highlight a crucial aspect of its ongoing advantage. Again, because of email's ubiquity, we tend to overlook the significance of this global system for enabling distributed, nonproprietary, and cross-platform communication. Such views often appeared in the surveys I ran. As one respondent put it, "Email is necessary because if someone doesn't use the social media medium that you use and if that someone forwards a message to you from that medium, if you didn't have email then you would not know." For Rob Mueller, of the email provider Fastmail I discussed in chapter 3, this feature is central to the role of email in the contemporary digital landscape seemingly dominated by social media. As he explains:

> The biggest area I would say where email is still special is that it's one of the few open and decentralized communication standards that are around on the internet, when you think about pretty much everything else. Like Facebook is all within Facebook, and everything is about keeping it within Facebook, and nothing shall exit Facebook. … And so Facebook [is] very keen on "you must use the Facebook App, you must use the Facebook website." … The point is more that they're very much sly [with the] keep you "in" kind of thing. Where email has and still is this global decentralized system so any email system can talk to any other system. And there's very few of them left. There are some others, insofar as the web is kind of [like] that. … Any browser can talk to any server but you don't have to go via Google, you don't have to go via Facebook. Anyone can talk to anyone.

Furthermore, in an era of increased visibility, email offers a little pocket of control for a number of my respondents. As one participant explained, "Email is a great way to communicate instead of Facebook if you don't want all your FB friends to know you are online chatting to someone because then everyone wants to talk." A study by Kenzie Burchell, who found that email represents a way to manage and limit exposure on social networking sites, echoes these attitudes.[8]

Perhaps because of this interoperability and relatively open standards, email is emerging as a crucial issue for digital inclusion, ensuring older citizens remain engaged in using tools for public and personal communication. Explaining the ongoing utility of email, one person answered the survey by saying:

> In terms of my use of e-mail, please note I am a low-tech person. I am conservative in terms of adapting to new technologies, driven in part by my age (70). And my personal style is NOT to share every detail of my life with half the world; thus I have no Facebook account.

Such findings are not surprising. After all, tech journalists have been writing for some time about the flight from email by young people. As one writer for *Gizmodo* puts it, "Email is the new generation gap."[9] That conviction is shared by a *TechCrunch* writer who explains, "The greatest threat to email is millennials."[10] According to a *Gizmodo* survey of 3,670 readers, "people over 40 are part of an email-centric generation." The survey revealed "a remarkable difference between people over 41, who say they often talk on email more than they do in real life, and younger people who love in-person meetings and use a variety of apps," which made *Gizmodo* wonder: "Is this our first digital generation gap?" Based on responses to the question about how people communicate with friends, the survey found that 74 percent of those aged between 41 and 70 chose email contrasted with only 52 percent of respondents under 40 who were "far more likely to talk using Facebook (69 per cent), texting (83 per cent), and the phone (61 per cent)."[11]

I have no problem with these *Gizmodo* figures per se, but we should always be wary of how easily the "death of email narrative" is fuelled by distaste for any aging media form and its consumers. As I explained in the introduction, quite often stories of media obsolescence and the incessant celebration of the new as "progress" are underpinned by a desire to escape material and (often) corporeal constraints. The past and its inefficient technologies, we are continually warned, will limit or even prevent commerce and communication. It is obviously the briefest of steps to conflate such a technological utopia with a fear of human ageing.

But I'm not quite finished yet with the *Gizmodo* study. Echoing beneath its figures (helpfully illustrated with an image of Tom Hanks at his keyboard from *You've Got Mail!*) is an odd contradiction. Summing up its findings the piece explains:

> It's clear that people over 40 are part of an email-centric generation.... Over 40s say they use email more than in-person meetings at work. That last statistic is rather poignant. This is a generation of people whose days are ruled by a slow, annoying, spam-ridden app. No wonder this generation has churned out thousands of books, movies, and other stories about how technology is eroding relationships and mangling our minds.... People below the age of 40, despite growing up with a mature internet and tons of mobile devices, say they communicate in person more than their elders. Of course, they are also online all the time. And there are definite generational differences within this group as well. The big gap between 30-somethings and 20-somethings seems to be Snapchat: only 12 per cent of 30-somethings talk to friends with Snapchat, while 32 per cent of 20-somethings do.[12]

So, on the one hand those under 40 are praised for their preference of face-to-face communication over the technologically mediated choices of their older counterparts, but on the other hand these "second-generation internet users" (between the ages of 21 and 40) are always online. The pivot point here seems to be email. Face-to-face authenticity is contrasted only with email rather than against other mediated forms such as Facebook or Snapchat. I do not want to belabor the point because this survey is "just" part of a tech blog. But it is instructional for thinking through how email is understood, particularly as it now seems to be used to represent old media.

Archives

As I explored in chapter 7, the postal system has been a source of inspiration to creative practitioners for centuries. While people were busy writing fiction composed of letters—Samuel Richardson's *Pamela* (published in 1740) and *The Sorrows of Young Werther* by Johann Wolfgang von Goethe (published in 1799), to name two—real letters were also being collected and published. So much of our history is informed by what we glean from the letters of politicians, actors, moguls, and activists that museum and curatorial policy has been developed to ensure the safe archival of these public-sphere resources. In cultural heritage organizations, a vital aspect of historical data collection comprises diaries, correspondence, personal ephemera, and other "grey literature." The same situation is now occurring to our current sphere of literary production with the MuseWeb (formerly Museums and the Web) project asking, "Who is generating the 'important" emails?'"[13]

Throughout the book, email archiving has underpinned many of the discussions from the history of the RFCs, itself a rich repository of internet history, to decisions by email group moderators about whether to open their archives to the public. And in chapter 6, I tried to come to grips with what it's been like "living in the archive" for those former Enron employees. Moreover, the Enron Corpus itself has provided a significant test bed for data science research attempting to develop new tools of email search, visualization, and preservation.[14] Now I want to draw some of these threads together by focusing specifically on cultural heritage policies, organizations, and tools that are attempting to collect and make available email communication.

The starting point for many discussions about how best to approach the preservation of email is the history of paper correspondence. Letters, like emails, can provide a window to the past for historical research on politics, economics, arts, education, and technology. Cultural heritage organizations have long included letters in their acquisitions policy and continue to develop sophisticated tools for storing, annotation, and searching. As evidence for the ongoing cultural fascination with historical letters is the crowd-funded project *Letters of Note* by Sean Usher. Originally established as a website for gathering letters by famous people, the project migrated to a highly successful book series made possible by the collective delight of a public intrigued by what letters can reveal. Usher runs the site offering a comprehensive set of navigation features including format (letter, memo, telegram, fax); date of correspondence (pre-1600 to post-2000); and categories (such as advice, anger, apology, fan letters, music, politics, racism, science, sexism, sport, Star Trek, technology, and war).[15] More formally are the correspondence holdings maintained by museums and libraries such as the astounding "Browning Letters" at Baylor University, which houses one of the most significant archives of correspondence exchanged between the Victorian poets Robert Browning and Elizabeth Barrett Browning.[16]

Although it's possible to establish the value of and necessity for archiving email by looking at the history of correspondence, this does not seem to be translating easily into practice. As William Vinh-Doyle argues, while "most archives organisations have invested resources to collect traditional analogue correspondence, the same attention to modern forms of communication, such as email, has not emerged."[17] Vinh-Doyle points out a key problem facing contemporary archivists: policy directives are often at odds with email users' actual every day habits. Corporate and government organizations may try to implement policy requiring its employees to follow record compliance directives, where mailboxes should be treated "in the same way you would manage records in other forms, such as paper," but unfortunately people simply do not comply. Years of document management research show that's not how people use email, either institutionally or personally. When email records are donated to cultural heritage institutions, become a focus for Freedom of Information Act requests, or are the target of legal discovery processes, their unruly nature becomes a problem. As he puts it, with the "failure of email management to gain ground in the last 20 years, archivists can expect to receive an exponential growth of unmanaged email from users, including government employees." To give some context to the scale of the problem Vinh-Doyle reports that his own organization, the Provincial Archives of New Brunswick (in Canada), has acquired "approximately 2 million email records from various senior officials" and experienced an "increase in the number of requests for information relating to these email messages from researchers and legal firms."[18] When these records arrive they are often unwieldy to navigate or access due to software obsolescence. Moreover, inadequate tools employed to open digital files acquired by the collection may often risk permanently altering or corrupting the original document. In the case of notoriously unstable .pst files (Personal Storage Table, which is Microsoft's email storage format), Vinh-Doyle explains how his organization struggled to view the records without risking "catastrophic loss." In order to address such challenges Vinh-Doyle recommends that the museum and library sectors begin to adopt "digital forensics" as a new mode of record management. Such recommendations pick up on what I mentioned earlier in relation to comparisons drawn with managing paper archives, and unfortunately the correlation is often not workable. As Andrew Waugh argues, the issues now facing cultural heritage organizations attempting to archive email communication is that "we are still trying to manage records as if they were mid-twentieth-century paper files."[19]

Digital forensics has become a significant response to issues facing the data archival sector. In their 2010 report "Digital Forensics and Born-Digital Content in Cultural Heritage Collections," Matthew Kirschenbaum, Richard Ovenden, and Gabriela Redwine define this approach as an "applied field originating in law enforcement, computer security, and national defense...concerned with discovering, authenticating, and analyzing data in digital formats to the standard of admissibility in a legal setting." As the authors further claim:

> Methods and tools developed by forensics experts represent a novel approach to key issues and challenges in the archives and curatorial community. Libraries, special collections, and other collecting institutions increasingly receive computer storage media (and sometimes entire computers) as part of their acquisition of "papers" from contemporary artists, writers, musicians, government officials, politicians, scholars, scientists, and other public figures. Smart phones, e-book readers, and other data-rich devices will surely follow.[20]

Cost is a serious problem facing institutions that wish to adopt a digital forensics regime. Most of the systems developed for forensic investigation are proprietary and thus attract high licensing fees, like the well-known EnCase software suite. In 2011 scoping began on producing an open source, web-accessed product resulting in the BitCurator Project. The initiative aims to "develop software for collecting institutions to extract, analyze, and produce reports on features of interest in text extracted from born-digital materials contained in collections."[21] Digital forensics is not the only approach being undertaken in cultural heritage contexts, and it is also not without its detractors. Surfacing deleted files and correspondence containing sensitive items or material not intended for publication by its authors is an ongoing issue for such initiatives.[22]

A number of large-scale, internationally focused projects have been launched over the last decade to address the specific issues involved in the digital preservation of email. As mentioned, at the MuseWeb Conference in 2014 a working party formed to explore the "range of knowledge management, conservation, and preservation issues that email presents for art museums" with a view to identify or create archiving tools to help the cultural sector "preserve significant emails that are vital to an institution's compliance with local, state, and federal laws, as well as with the specific legal standards governing trust responsibilities."[23] Prior to this initiative the Collaborative Electronic Records Project (CERP), a joint partnership between the Rockefeller Archive Center and the Smithsonian Institution Archives, ran from 2003 to 2008. CERP had a focus on email archival practice because the "complexity of email records poses a special preservation challenge." In particular the "basic functionality of email 'threads' was at the heart of this challenge. An email can contain several emails either in the body of the uppermost email or as attachments."[24] In November 2016 the Task Force on Technical Approaches for Email Archives was announced, jointly funded by the Andrew W. Mellon Foundation and the UK-based, not-for-profit Digital Preservation Coalition (DPC). At its heart, the task force recognizes that "email has remained resistant to a variety of efforts at preservation and is currently not systematically acquired by most archives and libraries." Underpinning the challenge is the fact that "email is no one thing, but rather a complicated interaction of technical subsystems for composition, transport, viewing, and storage," which means preserving email "involves multiple processes including acquisition and appraisal of collections, processing records, meeting privacy and legal considerations, preserving messages and attachments, and facilitating access.'[25]

What the ongoing studies on email archival practice reveal is the crucial role being played by everyday context. Years of research show that in general people do not follow the directives of their employers to apply record-management policies to their inboxes. In part this is because the purposes of email bleeds into other applications such as calendar management, attachments, and meeting minute-keeping. So, for example, where an organization policy may direct its staff to file its records the way they would for paper documentation, it is sometimes difficult to carry out because a particular record could stretch over many different digital environments. The minutes of a meeting can reside in the calendar appointment, as an attachment or embedded within the body of an email. So we now need urgently to devote our time to the vast archives of popular, social, personal, and political emails that are amassing within and constituting the contours of our public sphere. We should take as inspiration the many imaginative and comprehensive projects that continue to document the epistolary archive: letters of famous people and everyday contexts are mapped and catalogued across the world by cultural heritage organizations and dedicated writers. We need to foster the same cultural appetite for email.

The Killer App

Right from the beginning email was overlooked. As Janet Abbate and others have shown, the emergence and striking success of email in the 1970s and 1980s came as an almost total shock to those designing the early technologies. Across the literature there seems to be consensus that this was due to miscalculating two important factors: first, the power of users in shaping and creating their own technological environments; and second, the significant role played by communication in the lives of those early users and builders of the network. This observation probably seems like a no-brainer to us now. Of course people want to use computers to communicate. But that is often precisely the manner in which media communications develop, in uneven and surprising ways. As Abbate explains:

> Since the original view of the network planners was that "resources" meant massive, expensive pieces of hardware or huge databases, they did not anticipate that people would turn out to be the network's most valued resources. Network users challenged the initial assumptions, voting with their packets by sending a huge volume of electronic mail but making relatively little use of remote hardware and software.... Increasingly, people within and outside the ARPA community would come to see the ARPANET not as a computing system but rather as a communications system.[26]

By the 1990s the widespread popularity of email was sealed with its description as the killer app of the internet. A so-called killer application refers to an innovation, usually involving digital technologies, that encourages adoption of a broader system or product. For example, an attractive piece of software may prompt consumers to buy the hardware on which it runs. So it was with the internet, which was suddenly made desirable; people got dial-up so they could use email. With email came addresses; consumers were now electronically identifiable and locatable in ways that had not happened at the same scale before. As Donna Haraway recognized very early on, the email address represents a new way to approach identity.[27] Part of the reason that email earned its killer-app status was because it humanized and personalized the wider, amorphous, and to many, inexplicable, technology of the internet. Email has always been graspable on that scale. Looking back, a further reason to account for its near ubiquitous global reach is the impact brought by an asynchronous communication system. Although, again, our contemporary digital landscape brims with options for asynchronous communication where parties do not have to occupy the same geotemporal location, in the 2000s this was not the case. Along with the advent of voicemail and fax, email presented one of the first worldwide, easy to use, cheap and accessible methods for conducting business and pleasure on our own spatiotemporal terms.

So much for its introduction. What then explains its remarkable persistence and endurance? Obviously there's no one, single answer to such a question. My interest has never really been in the expansive, decisive, epochal pronouncements of the sort beloved by those big-picture theorists who select a technology to stand for monumental shifts in the material and social world. I'm not saying these aren't incredibly valuable for moving scholarship along, but it is not the method I have employed. Nonetheless, in addition to the features already discussed above—a relatively open global communication system based on interoperability, nonproprietary standards, and a vendor-neutral approach—it could actually come down to the way it is bemoaned. While it might sound fanciful to suggest that email is successful because everyone hates it, think about how mobile devices have now reached almost saturation level and what happens at that point are the complaints about its ubiquity. As Peter Ling has so insightfully shown about the mobile phone, there is an interesting link between the almost complete diffusion of a technology in a culture, its "disappearance" as he puts it, and the way it is then treated with disdain and annoyance.[28] Similarly, I am always struck by the way students often love to talk about their slavish adoration to and dependence on their phones or social media apps in ways that are not straightforward. In mapping the ways in which we take email for granted I hope I have also unearthed its compelling stories.

Notes

Introduction

1. Lucy Mangan, "When the French Clock Off at 6pm, They Really Mean It," *Guardian*, April 10, 2014, http://www.theguardian.com/money/shortcuts/2014/apr/09/french-6pm-labour-agreement-work-emails-out-of-office.

2. Simon Kelner, "French Workers Now Have Legal Right Not to Be Contacted after They Leave the Office," *Independent*, April 9, 2014, http://www.independent.co.uk/news/world/europe/french-workers-now-have-legal-right-not-to-be-contacted-after-they-leave-the-office-9249506.html.

3. Mangan, "When the French Clock Off."

4. Allison P Davis, "Two French Unions Ban Checking Work Emails after 6 p.m," *NYMAG.com*, April 10, 2014, http://nymag.com/thecut/2014/04/france-bans-checking-work-emails-after-6-pm.html.

5. John Johnson, "France Bans Work E-mail After 6 p.m." *USA Today*, April 11, 2014, http://www.usatoday.com/story/money/business/2014/04/11/newser-france-work-email-ban/7592125/.

6. "French Tech Workers: No E-Mails After 6," *FOXBusiness*, April 10, 2014, http://www.foxbusiness.com/personal-finance/2014/04/10/french-tech-workers-no-e-mails-after-6/.

7. Vivienne Walt, "France's 'Right to Disconnect' Law Isn't All It's Cracked Up to Be," *Time*, January 5, 2017, http://time.com/4622095/france-right-to-disconnect-email-work/.

8. Peter Brooks, *Reading for the Plot: Design and Intention in Narrative* (Cambridge, MA: Harvard University Press, 1984).

9. Mahesh Sharma, "Kill Off Email to Boost Productivity," *The Age*, April 11, 2014, http://www.theage.com.au/digital-life/digital-life-news/kill-off-email-to-boost-productivity-20140411-zqtik.html.

10. Marcus Wohlsen, "The Next Big Thing You Missed: Email's About to Die, Argues Facebook Co-founder," *Wired*, January 21, 2014, https://www.wired.com/2014/01/next-big-thing-missed-facebook-co-founder-says-email/.

11. John McCarthy, "Networks Considered Harmful for Electronic Mail," *Communications of the ACM* 32, no.12 (1989): 1389–1390.

12. Federal Trade Commission, "Spam," http://www.consumer.ftc.gov/articles/0038-spam.

13. Patrick Flanagan, "Spam: Not Just for Breakfast Anymore: Unsolicited Email in the Business Environment," conference paper, 4th Annual Ethics and Technology Conference, June 4–5, 1999, http://www.bc.edu/bc_org/avp/law/st_org/iptf/commentary/content/1999060511.html.

14. Kevin Werbach, "Death by Spam: The E-mail You Know and Love Is about to Vanish," *Slate*, November 18, 2002, http://www.slate.com/articles/technology/webhead/2002/11/death_by_spam.html.

15. Rita-Marie Conrad, "Save Yourself from Drowning in Online Interaction," Designing for Learning, 1999, http://www.designingforlearning.info/services/writing/conrad.htm.

16. Jordan Furlong, "The Last Days of E-mail," *Law 21* (blog), February 29, 2008, http://www.law21.ca/2008/02/the-last-days-of-e-mail/.

17. "Atos Boss Thierry Breton Defends His Internal Email Ban," *BBC News*, December 6, 2011, http://www.bbc.com/news/technology-16055310.

18. Thierry Breton, Chairman and CEO of Atos, Atos website, http://atos.net/en-us/home/we-are/zero-email.html.

19. Martin Bosworth, "The Death of E-Mail?," *Consumer Affairs*, July 27, 2006, https://www.consumeraffairs.com/news04/2006/07/email_fading.html.

20. Chad Lorenz, "The Death of E-Mail: Teenagers are Abandoning Their Yahoo! and Hotmail Accounts. Do the Rest of Us Have To?," *Slate*, November 14, 2007, http://www.slate.com/articles/technology/technology/2007/11/the_death_of_email.html.

21. Mike Dover, "More News about the Death of Email," *Wikinomics*, June 18, 2008, http://www.wikinomics.com/blog/index.php/2008/06/18/more-news-about-the-death-of-email/.

22. Tim Young, "Social Networks Spur the Demise of Email in the Workplace," *Socialcast Blog*, June 7, 2010, http://blog.socialcast.com/social-networks-spur-the-demise-of-email-in-the-workplace/.

23. Charles R. Acland, ed., *Residual Media* (Minneapolis: University of Minnesota Press, 2007).

24. Frederich Kittler, *Gramophone, Film, Typewriter* (Stanford, CA: Stanford University Press, 1999); Siegfried Zielinski, *Deep Time of the Media* (Cambridge, MA: MIT Press, 2006); Jussi Parikka *What Is Media Archaeology*? (Cambridge: Polity Press: 2012).

25. Acland, *Residual Media*, xviii.

26. Ryan Holmes, "Email Is the New Pony Express—And It's Time to Put It Down," *Fast Company*, October 16, 2012: http://www.fastcompany.com/3002170/email-new-pony-express-and-its-time-put-it-down.

27. The Radicati Group, "Email Statistics Report, 2020–2024," March 2, 2020, https://www.radicati.com/?p=16510.

28. Statista, "Number of Active Gmail Users … ," https://www.statista.com/statistics/432390/active-gmail-users/.

29. Statista, "Most Popular Mobile Internet Activities … ," https://www.statista.com/statistics/249761/most-popular-activities-carried-out-on-mobile-internet-devices/.

30. Maurice V. Wilkes, "Networks, Email and Fax," *Communications of the ACM* 33, no. 6 (1990): 631–633.

31. Lee Sproull and Sara Kiesler, "Reducing Social Context Cues: Electronic Mail in Organizational Communication," *Management Science* 32, no. 11 (1986): 1492–1512.

32. For literature reviews on flaming scholarship, see Martin Lea et al., "'Flaming' in Computer-Mediated Communication: Observations, Explanations, Implications," *Contexts of Computer-Mediated Communication*, ed. Martin Lea (Hertfordshire, UK: Harvester, 1992) 94–95; Philip A. Thompsen, "What's Fueling the Flames in Cyberspace? A Social Influence Model," *Communication and Cyberspace: Social Interaction in an Electronic Environment*, ed. Lance Strate, Ron L. Jacobson, and Stephanie Gibson (Cresskill, NJ: Hampton, 1996); Anna K. Turnage, "Email Flaming Behaviors and Organizational Conflict," *Journal of Computer-Mediated Communication* 13 (2008): 43–59; Esther Milne, *Letters Postcards Email; Technologies of Presence* (London: Routledge, 2010), 170–172.

33. See, for example, Joanne Goode and Maggie Johnson, "Putting out the Flames: The Etiquette and Law of E-Mail," *ONLINE: The Magazine of Online Information Systems* 15, no. 6 (1991): 61–65; Philip A. Thompsen and Davis A. Foulger, "Effects of Pictographs and Quoting on Flaming in Electronic Mail," *Computers in Human Behavior* 12, no. 2 (1996): 225–243; Edward A. Mabry, "Framing Flames: The Structure of Argumentative Messages on the Net," *Journal of Computer-Mediated Communication* 2, no. 4 (1997); Steven S. Vrooman, "The Art of Invective: Performing Identity in Cyberspace," *New Media & Society* 4, no.1 (2002): 51–70; William B. Millard, "I Flamed Freud: A Case Study in Teletextual Incendarism," in *Internet Culture*, ed. David Porter (New York: Routledge, 1997), 145–159.

34. See, for example, "Trolls and The Negative Space of the Internet," *Fibreculture Journal* 22 (2013), special issue edited by Jason Wilson, Glen Fuller, and Christian McCrea, http://twentytwo.fibreculturejournal.org/.

35. Joseph Schmitz and Janet Fulk, "Organizational Colleagues, Media Richness, and Electronic Mail," *Communication Research* 18, no. 4 (1991): 487–523.

36. James Keaten and Lynne Kelly, "'Re: We Really Need to Talk': Affect for Communication Channels, Competence, and Fear of Negative Evaluation," *Communication Quarterly* 56, no.4 (2008): 407–426; Yusun Jung and Kalle Lyytinen, "Towards an Ecological Account of Media Choice: A Case Study on Pluralistic Reasoning While Choosing Email," *Information Systems Journal* 24, no. 3 (2014): 271–293.

37. See, for example, Denise E. Murray, "CMC: A Report on the Nature and Evolution of On-Line E-Messages," *English Today* 6, no. 3, (1990): 42–46; Naomi S. Baron, "Why Email Looks Like Speech: Proofreading, Pedagogy and Public Face," in *New Media Language*, ed. Jean Aitchison and Diana M. Lewis (London, Routledge: 2003), 85–94; Whitney Bolton, "CMC and Email: Casting a Wider Net," *English Today* 7, no. 4 (1991): 34–38; Fraida Dubin, "Checking Out E-mail and the Fax," *English Today* 7, no. 1 (1991): 47–51; Andrea Ovans, "Can E-mail Deliver the Message?," *Datamation* 38, no. 11 (1992); Simeon J. Yates, "Oral and Written Linguistic Aspects of Computer Conferencing: A Corpus Based Study," in *Computer-Mediated Communication: Linguistic, Social and Cross-Cultural Perspectives*, ed. Susan C. Herring (Amsterdam: Benjamins, 1996) 29–46.

38. See, for example, Patricia A. Merrier and Ruthann Dirks, "Student Attitudes Toward Written Oral and E-mail Communication," *Business Communication Quarterly* 60, no. 2 (1997): 89–99; San Bolkan and Jennifer Linn Holmgren, "You Are Such a Great Teacher and I Hate to Bother You But ...": Instructors' Perceptions of Students and Their Use of Email Messages with Varying Politeness Strategies," *Communication Education* 61, no. 3 (2012): 253–270; Wuhan Zhu, "Polite Requestive Strategies in Emails: An Investigation of Pragmatic Competence of Chinese EFL Learners," *RELC Journal* 43 (2012): 217–238; and Hee Sun Park, Hye Eun Lee, Jeong An Song, "'I Am Sorry to Send You SPAM': Cross-Cultural Differences in Use of Apologies in Email Advertising in Korea and US," *Human Communication Research* 31, no. 3 (2005): 365–398.

39. Carmen Frehner, *Email—SMS—MMS: The Linguistic Creativity of Asynchronous Discourse in the New Media Age* (New York: Linguistic Insights, 2008); Naomi S. Baron, *Alphabet to Email: How Written English Evolved and Where It's Heading* (London: Routledge, 2000); David Crystal, *Internet Linguistics: A Student Guide* (London: Routledge, 2011).

40. Brenda Danet, *Cyberpl@y: Communicating Online* (London: Bloomsbury, 2001); Sunka Simon, *Mail-Orders: The Fiction of Letters in Postmodern Culture* (Albany: State University of New York, 2002); Emma Rooksby, *E-Mail and Ethics: Style and Ethical Relations in Computer-Mediated Communication* (London: Routledge, 2002).

41. Mark Mabrito, "The E-Mail Discussion Group: An Opportunity for Discourse Analysis," *Business Communication Quarterly* 58, no. 2 (1995): 10–11; Karen Nantz and Cynthia Drexel, "Incorporating Electronic Mail into the Business Communication Course," *Business Communication Quarterly* 58, no. 3 (1995): 45–51; Chad Hilton and Naoki Kameda, "E-Mail and the Internet as International Business Communication Teaching and Research Tools—A Case Study," *Journal of Education for Business* 74, no. 3 (1999): 181–186.

42. Nicole Cásarez, "Electronic Mail and Employee Relations: Why Privacy Must Be Considered," *Public Relations Quarterly* 37, no. 2 (1992): 37–40.

43. Margaret Steen, "Legal Pitfalls of E-mail," *InfoWorld* 21, no. 27 (1999): 65–66.

44. Marian K. Riedy and Joseph H. Wen, "Electronic Surveillance of Internet Access in the American Workplace: Implications for Management," *Information & Communications Technology Law* 19, no.1 (2010): 87–99; *Fourteenth Annual Report of the Article 29 Working Party on Data Protection*, Luxembourg: Publications Office of the European Union, 2013, http://ec.europa.eu/justice/data-protection/article-29/documentation/annual-report/files/14th_annual_report_en.pdf.

45. William Smith and Filiz Tabak, "Monitoring Employee E-mails: Is There Any Room for Privacy?" *Academy of Management Perspectives* 23, no. 4 (2009): 33–48.

46. In addition to those references covered in the text see Douglas Shaller, "E-mail, the Internet, and Other Legal and Ethical NIGHTMARES," *Strategic Finance* 82, no. 2 (2000): 48–52, and Micah Echols, "Striking a Balance Between Employer Business Interests and Employee Privacy: Using *Respondeat Superior* to Justify the Monitoring of Web-Based, Personal Electronic Mail Accounts of Employees in the Workplace," *Computer Law Review and Technology Journal* 7 (2003): 273–300.

47. Jan Samoriski and John Huffman, "Electronic Mail, Privacy, and the Electronic Communications Privacy Act of 1986: Technology in Search of Law," *Journal of Broadcasting & Electronic Media* 40, no. 1 (1996): 60.

48. Thomas J. Hodson, Fred Englander, and Valerie Englander, "Ethical, Legal, and Economic Aspects of Employer Monitoring of Employee Electronic Mail," *Journal of Business Ethics* 19 (1999): 99–108; see also Carl Botan, "Communication Work and Electronic Surveillance: A Model for Predicting Panoptic Effects," *Communication Monographs* 63, no. 4 (1996): 293–313; Jason L. Snyder, "E-mail Privacy in the Workplace: A Boundary Regulation Perspective," *Journal of Business Communication* 47, no. 3 (2010): 266–294.

49. See, for example, Ian Brodie, *Email Persuasion: Captivate and Engage Your Audience, Build Authority and Generate More Sales With Email Marketing* (Myrtle Beach, SC: Rainmaker, 2013); D.J. Waldow and Jason Fells, *The Rebel's Guide to Email Marketing: Grow Your List, Break the Rules, and Win* (Indianapolis: Que Publishing: 2012).

50. David Shipley, *SEND: Why People Email So Badly and How to Do It Better* (Visalia, CA: Vintage, 2007, 2010); Jim McCullen, *Control Your Day: A New Approach to Email*

and Time Management Using Microsoft® Outlook and the concepts of Getting Things Done® (Brookfield, WI: Stone River, 2013).

51. See, for example, Joseph E. Phelps, Regina Lewis, Lynne Mobilio, et al., "Viral Marketing or Electronic Word-of-Mouth Advertising: Examining Consumer Responses and Motivations to Pass Along Email," *Journal of Advertising Research* 44, no. 12 (2004): 333–348; Navdeep S. Sahni, S. Christian Wheeler, and Pradeep Chintagunta, "Personalization in Email Marketing: The Role of Non Informative Advertising Content," *Marketing Science* 37, no. 2 (2018): 236–258.

52. The major works in the field of internet history are: Peter Salus, *Casting the Net: From ARPANET to Internet and Beyond…* (Boston: Addison-Wesley, 1995); Janet Abbate, *Inventing the Internet* (Cambridge, MA: MIT Press, 1999); Katie Hafner and Matthew Lyon, *Where Wizards Stay Up Late* (New York: Simon & Schuster, 1996); Michael Hauben and Ronda Hauben, *Netizens: On the History and Impact of Usenet and the Internet* (Los Alamitos, CA: Computer Science Press, 1997); Arthur Norberg and Judy O'Neill Freedman, *Transforming Computer Technology: Information Processing for the Pentagon, 1962–1986* (Baltimore: Johns Hopkins University Press, 1996). See also: Craig Partridge, "The Technical Development of Internet Email," *IEEE Annals of the History of Computing* (2008) 3–29.

53. John Carey and Martin C.J. Elton, "The Other Path to the Web: The Forgotten Role of Videotex and Other Early Online Services," *New Media Society* (2009): 241–260.

54. Melanie Swalwell, "Questions about the Usefulness of Microcomputers in 1980s Australia," *Media International Australia* 143 (2012): 63–77.

55. Tim Highfield and Axel Bruns, "Confrontation and Cooptation: A Brief History of Australian Political Blogs," *Media International Australia* 143 (2012): 89–98.

56. Vivienne Waller, "'This Big Hi-tech Thing': Gender and the Internet at Home in the 1990s," *Media International Australia* 143 (2012): 78–88.

57. Gerard Goggin and Mark McLelland, *The Routledge Companion to Global Internet Histories* (London: Routledge, 2017).

58. Juss Parikka and Tony Sampson, eds., *The Spam Book: On Viruses, Porn, and Other Anomalies from the Dark Side of Digital Culture* (Cresskill, NJ: Hampton Press, 2009); Finn Brunton, *Spam: A Shadow History of the Internet* (Cambridge, MA: MIT Press, 2013)

59. The literature review was conducted across the major international research databases of Scopus, Sage, EbscoHost, JSTOR, Gale, Academic Search Complete, Business Source Complete, and Communication & Mass Media Complete for the period from 1988 to 2018. Single journal searches were also run for leading media, communication, and cultural studies journals. Email and its variant spelling were searched in title, keywords, abstract, and full text. To count as a "substantial exploration of email" the term needed to appear in the title or keyword fields. When email appears

only in the abstract it is often merely a list of exemplary communication technologies: "SMS, email, social media," for example, and is therefore not indicative of a comprehensive analysis of email. But in some cases the appearance of email in the abstract *did* indicate substantial investigation and so these have been included in the tallies. In other cases, email did not appear in title, abstract, or keyword but turned out to be a substantial treatment of email and so was included. To provide a few concrete examples: in the internet-themed issue of *Media International Australia*, email appears only about 8 times and then as part of a generalized list of online platforms and services. *New Media and Society* has published 23 papers substantially about email, the *Journal of Communication* has published 6; in *Convergence* there have been 11; *Media, Culture and Society* has published 3; *The European Journal of Cultural Studies* has featured 5 articles about email and *Cultural Studies* has published 1 article substantially devoted to email. In contrast, there have been 75 papers published in *Computers in Human Behavior* and 129 papers devoted to email published in *Communications of the ACM*. One does need to be slightly wary of these figures. Limitations include the different publication periods for the journals. ACM has been in existence since the 1970s and has a different publishing remit than, say, *New Media and Society* so that the former accepts for publication a wider set of contributions. Also, journals vary on number of issues per year so obviously that can affect results. Finally, there are variances between indexing databases in terms of abstract or keyword criteria (not all journals surveyed contained article abstracts).

60. Christian Reutera, Thomas Ludwiga, Marc-André Kaufhold, and Thomas Spielhofer, "Emergency Services' Attitudes towards Social Media: A Quantitative and Qualitative Survey Across Europe," *International Journal of Human-Computer Studies* 95 (November 2016): 96–111.

61. The surveys were anonymous so no citation details are provided either in text or as endnotes. Interview participants were given options of anonymity, pseudonymity, or to be identified. Again, no citation details are provided following their quotes but effort is made in the text to give contextual information of participants. If no endnote appears for a quote then it means the material was directly communicated to me during the research.

62. Donna Haraway, "Situated Knowledges: The Science Question in Feminism and the Privilege of Partial Perspective," *Feminist Studies* 14, no. 3 (1988): 575–599. See also: Stuart Hall, "Cultural Studies and its Theoretical Legacies," in *Cultural Studies*, ed. Lawrence Grossberg, Cary Nelson, Paula Treichler (London: Routledge, 1992), 277–294.

63. Fran Martin, "Introduction," in *Interpreting Everyday Culture*, ed. Fran Martin (London: Hodder Arnold, 2003), 4.

64. Ben Highmore, *Everyday Life and Cultural Theory* (New York: Taylor and Francis, 2002), 1.

65. Joe Moran, *Reading the Everyday* (London: Routledge, 2005), 12.

66. Judy Attfield, *Wild Things: The Material Culture of Everyday Life* (Oxford: Berg, 2005), 50.

67. Ben Highmore, "Georges Perec and the Significance of the Insignificant," in *The Afterlives of Georges Perec*, ed. Rowan Wilken and Justin Clemens (Edinburgh: Edinburgh University Press, 2017), 105–119.

68. Anthony McCosker and Rowan Wilken, "'Things That Should Be Short': Perec, Sei Shōnagon, Twitter, and the Uses of Banality," in *The Afterlives of Georges Perec*, ed. Rowan Wilken and Justin Clemens (Edinburgh: Edinburgh University Press, 2017), 138.

69. Highmore, "Georges Perec and the Significance of the Insignificant."

70. Melissa Gregg, *Work's Intimacy* (Cambridge: Polity, 2011), 16–17.

71. Alasdair Jones, "Everyday Without Exception? Making Space for the Exceptional in Contemporary Sociological Studies of Streetlife," *The Sociological Review*, 66, no. 5 (2018): 1000–1016.

72. Victoria Robinson, "Reconceptualising the Mundane and the Extraordinary: A Lens through Which to Explore Transformation within Women's Everyday Footwear Practices," *Sociology* 49, no. 5 (2015): 904.

73. Nick Couldry, *Media, Society, World: Social Theory and Digital Media Practice* (Cambridge: Polity, 2013), 16.

74. Nick Couldry and Andreas Hepp, *The Mediated Construction of Reality* (Cambridge: Polity, 2017), 54–55.

75. Mirca Madianou and Daniel Miller, "Polymedia: Towards a New Theory of Digital Media on Interpersonal Communication," *International Journal of Cultural Studies* 16, no. 2 (2013): 170.

76. Ilana Gershon, *The Breakup 2.0: Disconnecting over New Media* (New York: Cornell University Press, 2010), 3.

77. Highmore, *Everyday Life and Cultural Theory*, 115.

78. See, for example, Carole Anne Rivière and Christian Licoppe, "From Voice to Text: Continuity and Change in the Use of Mobile Phones in France and Japan," in *The Inside Text*, ed. Richard Harper, Leysia Ann Palen, and A. Taylor (Netherlands: Springer, 2006).

Chapter 1: The Origins of Email and Its Development

1. Tom Walker, "The Evolution of Printer Technology: Then and Now," *Cartridgesave.co.uk* (blog), April 15, 2008, http://www.cartridgesave.co.uk/news/the-evolution-of-printer-technology-then-and-now/.

2. Janet Abbate, *Inventing the Internet* (Cambridge: MIT Press, 1999), 106; Katie Hafner and Matthew Lyon, *Where Wizards Stay Up Late* (New York: Simon & Schuster, 1996),

191–192; Peter H. Salus, *Casting the Net: From ARPANET to Internet and Beyond* (Reading, MA: Addison-Wesley, 1995), 17.

3. Tomlinson, quoted by John Devlin, "Ray Tomlinson, the Inventor of Email and @ Usage, Dies at 74," Neowin.net website, https://www.neowin.net/news/ray-tomlinson-the-inventor-of-email-and--usage-dies-at-74/.

4. "Oral History of Raymond 'Ray' Tomlinson," interviewed by Marc Weber and Gardner Hendrie, June 5, 2009, Cambridge, Massachusetts, CHM Reference number: X5409.2009 2009, Computer History Museum.

5. "Raymond Tomlinson," Internet Hall of Fame, http://internethalloffame.org/inductees/raymond-tomlinson.

6. Tom Van Vleck, "The History of Electronic Mail," http://multicians.org/thvv/mail-history.html.

7. Tom Van Vleck, "Electronic Mail and Text Messaging in CTSS, 1965–1973," *IEEE Annals of the History of Computing* 34, no. 1 (2012): 4–6.

8. Craig Partridge, "The Technical Development of Internet Email," *IEEE Annals of the History of Computing* 30, no. 2 (2008): 4.

9. Hafner and Lyon, *Wizards Stay Up Late*, 192.

10. Partridge, "Technical Development of Internet Email," 6.

11. Partridge, 6.

12. Partridge, 7.

13. Partridge, 7.

14. Elizabeth Feinler, to SIGCIS members, February 29, 2012, "Who Invented Email?" http://www.sigcis.org/.

15. Partridge, "Technical Development of Internet Email," 7.

16. Hafner and Lyon, *Wizards Stay Up Late*, 204–205.

17. Partridge, "Technical Development of Internet Email," 7–8.

18. Thierry Bardini, *Bootstrapping: Douglas Engelbart, Coevolution, and the Origins of Personal Computing* (Stanford, CA: Stanford University Press, 2000), 187–193.

19. Bardini, *Bootstrapping*, 109.

20. Bardini, 109.

21. M. A. Padlipsky, "And They Argued All Night ..." in *The ARPANET Sourcebook: The Unpublished Foundations of the Internet*, ed. Peter H Salus (Charlottesville, VA: Peer-to-Peer Communications, 2008), 506. Emphasis in original.

22. Tom Van Vleck, "Dead Media Beat: The History of Electronic Mail," *Wired*, April 28, 2013, https://www.wired.com/2013/04/dead-media-beat-the-history-of-elec tronic-mail/.

23. Errol Morris, "Did My Brother Invent Email with Tom Van Vleck?" *New York Times*, June 19, 2011, http://opinionator.blogs.nytimes.com/2011/06/23/did-my -brother-invent-e-mail-with-tom-van-vleck-part-five.

24. Morris, "Did My Brother Invent Email with Tom Van Vleck?"

25. Eugene C. Johnson, Testimony, United States Postal Service Recommended Decision on Changes in the Classification Schedule (ECOM), Docket Number MC78–3, Postal Rate Commission, September 8, 1978, 8. My thanks to Leona Anasiewicz at the US Postal Regulatory Commission for supplying this documentation.

26. Devin Leonard, *Neither Snow nor Rain: A History of the United States Postal Service* (New York: Grove Press, 2016), 223.

27. National Research Council, *Review of Electronic Mail Service System for US Postal Service* (Washington, DC: National Academy Press, 1981), ix.

28. David Farber and Paul Baran, "The Convergence of Computing and Telecommunications Systems," *Science* 195, no. 4283 (March 18, 1977): 1166, https://doi.org /10.1126/science.195.4283.1166.

29. Ted Nelson, "Mail Chauvinism: The Magicians, the Snark and the Camel," *Creative Computing* 7, no. 11 (November 1981).

30. US Department of Justice (Antitrust Division), "Proposed Revisions in the Comprehensive Standards for Permissible Private Carriage of Letters," March 13, 1979, 16. My thanks to Leona Anasiewicz at the US Postal Regulatory Commission for supplying this documentation.

31. US Department of Justice, "Proposed Revisions," 34.

32. Leonard, *Neither Snow nor Rain*, 224.

33. National Research Council, *Review of Electronic Mail Service System for US Postal Service*, 17.

34. "Oral History of Raymond 'Ray' Tomlinson."

35. Thomas Haigh, "Protocols for Profit: Web and Email Technologies as Product and Infrastructure," in *The Internet and American Business*, ed. William Aspray and Paul E. Ceruzzi (Cambridge, MA: MIT Press, 2010), 105–158.

36. RFC Editor, "About Us," https://www.rfc-editor.org/about/.

37. Jon Postel, "Simple Mail Transfer Protocol," STD 10, RFC 821, https://doi.org/10 .17487/RFC0821, August 1982; John Klensin, "Simple Mail Transfer Protocol," RFC

2821, https://doi.org/10.17487/RFC2821, April 2001; John Klensin, "Simple Mail Transfer Protocol," RFC 5321, https://doi.org/10.17487/RFC5321, October 2008.

38. Andrew L. Russell, *Open Standards and the Digital Age: History, Ideology, and Networks* (New York: Cambridge University Press, 2014), 244.

39. Rogers and Kingsley quoted in Russell, *Open Standards and the Digital Age,* 246.

40. Juan D. Rogers and Gordon Kingsley, "Denying Public Value: The Role of the Public Sector in Accounts of the Development of the Internet," *Journal of Public Administration Research and Theory* 14, no. 3 (2004): 378.

41. Russell, *Open Standards and the Digital Age,* 16.

42. Russell, 232.

43. Andrew L Russell, "OSI: The Internet That Wasn't," *IEEE Spectrum,* July 30, 2013, http://spectrum.ieee.org/computing/networks/osi-the-internet-that-wasnt.

44. Craig Partridge, "The Technical Development of Internet Email," *IEEE Annals of the History of Computing* 30, no. 2 (2008): 22.

45. Kai Jakobs, "Why Then Did the X.400 E-mail Standard Fail? Reasons and Lessons to be Learned," *Journal of Information Technology* 28, no. 1 (2013): 63–73.

46. Nathaniel Borenstein, "Part # 146: We're Slaves to Our Attachments," *Mimecast* (blog), March 25, 2011, https://www.mimecast.com/blog/2011/03/part-146-were-slaves-to-our-attachments/.

47. Scott Bradner, "IETF Working Group Guidelines and Procedures," BCP 25, RFC 2418, https://doi.org/10.17487/RFC2418, September 1998.

48. Stephen Crocker, "Documentation Conventions," RFC 3, https://doi.org/10.17487/RFC0003, April 1969.

49. Jon Postel, "File Transfer Protocol Specification," RFC 765, https://doi.org/10.17487/RFC0765, June 1980.

50. Heather Flanagan, "The RFC Series and the Twenty-first Century," *IETF Journal,* July 2014, http://www.internetsociety.org/articles/rfc-series-and-twenty-first-century.

51. Sandra Braman, "The Interpenetration of Technical and Legal Decision-Making for the Internet," *Information, Communication & Society* 13, no. 3 (2010): 310.

52. Laura DeNardis, *Protocol Politics: The Globalization of Internet Governance,* (Cambridge, MA: MIT Press, 2009).

53. Alexander R. Galloway, *The Interface Effect* (Cambridge: Polity, 2012), 140.

54. Alexander R. Galloway, *Protocol: How Control Exists after Decentralization* (Cambridge, MA: MIT Press, 2004), 8.

55. Galloway, *Protocol*, 141.

56. Galloway, 121.

57. Ben Kafka, *The Demon of Writing: Powers and Failures of Paperwork* (Brooklyn, NY: Zone Books, 2012).

58. Sandra Braman, "Privacy by Design: Networked Computing, 1969–1979," *New Media & Society* 14, no. 5 (2012): 799.

59. Brian E. Carpenter and Craig Partridge, "Internet Requests for Comments (RFCs) as Scholarly Publications," *ACM SIGCOMM Computer Communication Review* 40, no. 1 (2010): 31.

60. Ken Pogran, John Vittal, Dave Crocker, and Austin Henderson, "Proposed Official Standard for the Format of ARPA Network Messages," RFC 724, May 12, 1977, https://doi.org/10.17487/RFC0724, 2–4.

61. Jon Postel, "Comments on RFC 724," Post to the MSGGRP discussion list, May 23, 1977.

62. Austin Henderson, "Re: Contents of Subject Fields," Post to the MSGGRP discussion list, April 25, 1977.

63. David H. Crocker, John J. Vittal, Kenneth T. Pogran, and D. Austin Henderson Jr., "Standard for the Format of ARPA Network Text Messages," RFC 733, November 21, 1977, https://doi.org/10.17487/RFC0733.

64. Hafner and Lyon, *Where Wizards Stay Up Late*, 200.

65. Partridge, "Technical Development of Internet Email," 9.

66. Pogran et al., "Proposed Official Standard," 7–8.

67. Pogran et al., 27.

68. P. Resnick, "Internet Message Format," RFC 5322, 10.17487/RFC5322 October 2008.

69. JoAnne Yates, *Control through Communication: The Rise of System in American Management*, (Baltimore: Johns Hopkins University Press, 1989).

70. Yates, *Control through Communication*, 95–97.

71. Nelson, "Mail Chauvinism: The Magicians, the Snark and the Camel."

72. Steve Crocker, "How the Internet Got Its Rules," *New York Times*, April 7, 2009, http://www.nytimes.com/2009/04/07/opinion/07crocker.html.

Chapter 2: "Inventing Email" and Doing Media History

1. Jay David Bolter and Richard Grusin, *Remediation: Understanding New Media* (Cambridge, MA: MIT Press, 1999); Jussi Parikka, *What Is Media Archaeology?* (Cambridge: Polity, 2012).

2. Leslie P. Michelson, "Foreword," in V.A. Shiva Ayyadurai, *The Email Revolution: Unleashing the Power to Connect* (New York: Allworth Press, 2013), xv.

3. Leslie P. Michelson, "The Invention of Email," http://historyofemail.com/invention-of-email.asp.

4. Leslie P. Michelson, "Recollections of a Mentor and Colleague of a 14-Year-Old, Who Invented Email in Newark, NJ," inventorofemail.com, n.d., http://www.inventorofemail.com/va_shiva_recollection_inventing_email_leslie_michelson.asp.

5. United States Copyright Office, "Computer Program for Electronic Mail System," 1981, http://cocatalog.loc.gov/cgi-bin/Pwebrecon.cgi?Search_Arg=Computer+program+for+electronic+mail+system&Search_Code=TALL&PID=LWXSm-GZV6_jJ2Ln9xo-BARvcPzkf_&SEQ=20161124215441&CNT=25&HIST=1.

6. "Livingston Student Designs Electronic Mail System," *West Essex Tribune*, October 30, 1980, https://vashiva.com/innovation/email/inv01.asp.

7. V.A. Shiva Ayyadurai, *The Email Revolution: Unleashing the Power to Connect* (New York: Allworth Press, 2013), 1.

8. Janelle Nanos, "Return to Sender," *Boston Magazine*, June 2012, http://www.bostonmagazine.com/2012/05/shiva-ayyaduri-email-us-postal-service/.

9. Doug Aamoth, "The Man Who Invented Email," *Time*, November 15, 2011, http://techland.time.com/2011/11/15/the-man-who-invented-email/5/.

10. "The Future of the Post Office," YouTube (video), 1:55:43, MIT Comparative Media Studies/Writing, posted April 20, 2016, https://www.youtube.com/watch?v=IcSQtTgWOQA; Rob McQueen, "The US Postal Service in the Technological Climate," *The Tech*, March 20, 2012, http://tech.mit.edu/V132/N13/email.html.

11. Joseph Stromberg, "A Piece of Email History Comes to the American History Museum," *Smithsonian Magazine*, February 22, 2012, http://www.smithsonianmag.com/smithsonian-institution/a-piece-of-email-history-comes-to-the-american-history-museum-105377178/.

12. Emi Kolawole, "Smithsonian Acquires Documents from Inventor of 'EMAIL' Program," *Washington Post*, February 17, 2012, https://www.washingtonpost.com/national/on-innovations/va-shivaayyadurai-inventor-of-e-mail-honored-by-smithsonian/2012/02/17/gIQA8gQhKR_story.html.

13. Thomas Haigh, to SIGCIS members, February 22, 2012, "Email Was Invented by a School Boy in 1978 Says Washington Post & Time Magazine," http://www.sigcis.org/.

14. Dave Crocker, "Re: V.A. Shiva Ayyadurai: Inventor of E-mail Honored by Smithsonian," February 18, 2012, https://www.ietf.org/mail-archive/web/ietf-smtp/current/msg00148.html.

15. "IEEE Internet Award Recipients," https://www.ieee.org/about/awards/bios/internet_recipients.html.

16. Emi Kolawole, "Smithsonian Acquires Documents."

17. Patrick B. Pexton, "Reader Meter: Who Really Invented E-mail?," *Washington Post*, February 24, 2012, https://www.washingtonpost.com/blogs/omblog/post/reader-meter-who-really-invented-e-mail/2012/02/24/gIQAHZugYR_blog.html.

18. Pexton, "Reader Meter."

19. Patrick B. Pexton, "Origins of E-mail: My Mea Culpa," *Washington Post*, March 1, 2012, https://www.washingtonpost.com/blogs/omblog/post/origins-of-e-mail-my-mea-culpa/2012/03/01/gIQAiOD5kR_blog.html.

20. Pexton, "Origins of E-mail."

21. "Statement from the National Museum of American History: Collection of Materials from V.A. Shiva Ayyadurai," February 23, 2012, http://americanhistory.si.edu/press/releases/statement-national-museum-american-history-collection-materials-va-shiva-ayyudurai.

22. Caleb Garling, "Who Invented Email? Just Ask…Noam Chomsky," *Wired*, June 16, 2012, https://www.wired.com/2012/06/noam-chomsky-email.

23. V.A. Shiva Ayyadurai, *The Email Revolution: Unleashing the Power to Connect* (New York: Allworth Press, 2013), 195–196.

24. Ayyadurai, *The Email Revolution*, 208.

25. Dave Farber to Interesting People members, February 20, 2012, "Re: [IP] V.A. Shiva Ayyadurai: Inventor of E-mail Honored by Smithsonian—The Washington Post," https://www.listbox.com/member/archive/247/2012/02/sort/thread/page/4/entry/4:99/20120219141237:ADCAD69A-5B2D-11E1-8876-F3FC0C963953.

26. Ayyadurai, *The Email Revolution*, 214.

27. Dave Walden, to SIGCIS members, March 24, 2016, "A Fresh Tissue of Lies," http://lists.sigcis.org/pipermail/members-sigcis.org/2016-March/000134.html.

28. Katie Hafner and Matthew Lyon, *Where Wizards Stay Up Late* (New York: Simon & Schuster, 1996), 200; Janet Abbate, *Inventing the Internet* (Cambridge: MIT Press, 1999), 70–73.

29. Mike Padlipsky, to [ih] internet-history members, April 14 2006, "[ih] internet-history Digest, Vol 14, Issue 5," http://mailman.postel.org/pipermail/internet-history/2006-April/000592.html.

30. Thomas Haigh, to SIGCIS members, "Did V.A. Shiva Ayyadurai Invent Email?," August 4, 2015, https://www.sigcis.org/ayyadurai#Sept2013.

31. William M. Bulkeley, "Echomail Provides an Answer For the Avalanche of E-Mail," *Wall Street Journal*, November 15, 2001, http://www.wsj.com/articles/SB1005776873756157960.

32. Finn Brunton, *Spam: A Shadow History of the Internet* (Cambridge, MA: MIT Press, 2013), 31.

33. The Computer History Museum, "EMMS; Electronic Mail and Message Systems; Electronic Mail and Micro Systems; Electronic Mail and Messaging Systems," catalog no. 102661013, 1977–2001, http://www.computerhistory.org/collections/catalog/102661013.

34. V.A. Shiva Ayyadurai, "Definition of Email," 2012–2019, http://www.inventorofemail.com/definition_of_email.asp.

35. Haigh, "Did V.A. Shiva Ayyadurai Invent Email?"

36. Einar Stefferud, "It Is Very Distressing that Present…Postal Management," post to the MSGGRP discussion list, May 10, 1978.

37. Geoff at SRI-AI to MSGGROUP members, January 5, 1976, "MSGGROUP# 248 ARPAnet mail the coming thing?," http://mercury.lcs.mit.edu/~jnc/tech/msggroup/msggroup0201-0300.txt.

38. William Uricchio, "History and Its Shadow: Thinking about the Contours of Absence," *Screen* 55, no. 1 (2014): 119–127.

39. V.A. Shiva, Twitter post, March 6, 2016, 6:39 p.m., https://twitter.com/va_shiva/status/706670699713335297.

40. V.A. Shiva, Twitter post, March 10, 2016, 7:45 p.m., https://twitter.com/va_shiva/status/708136758077030400.

41. V.A. Shiva, Twitter post, March 6, 2016, 3:40 p.m., https://twitter.com/va_shiva/status/706625616041406468.

42. Rebecca Greenfield, "The Other Tech Figure Who's Trying to Kill Gawker," *Bloomberg*, June 4, 2016, https://www.bloomberg.com/news/articles/2016-06-03/the-other-tech-figure-who-s-trying-to-kill-gawker.

43. Owen Thomas, "Peter Thiel Is Totally Gay, People," *Gawker* (blog), December 19, 2007, https://gawker.com/335894/peter-thiel-is-totally-gay-people; see also *Forbes*, May 18, 2009, http://www.forbes.com/pictures/emdh45egkhm/may-18-2009/.

44. Andrew Ross Sorkin, "Peter Thiel, Tech Billionaire, Reveals Secret War with Gawker," *New York Times*, May 25, 2016, http://www.nytimes.com/2016/05/26/business/dealbook/peter-thiel-tech-billionaire-reveals-secret-war-with-gawker.html.

45. Alexandros K. Antonioua and Dimitris Akrivosb, "Hulk Hogan and the Demise of Gawker Media: Wrestling with Problems of Celebrity Voyeurism, Newsworthiness, and Tabloidisation," *Journal of Media Law* 8, no. 2 (2016): 167.

46. Rebecca Greenfield, "Other Tech Figure," https://www.bloomberg.com/news/articles/2016-06-03/the-other-tech-figure-who-s-trying-to-kill-gawker.

47. "Shiva Ayyadurai v. Gawker Media," 1:16-cv-10853-JCB, May 10, 2016, 1–23, *Scribed Inc.*, https://www.scribd.com/doc/313925324/Shiva-Ayyadurai.

48. Erik Wemple, "Conde Nast Exec Story: Gawker is Keeping Its Sleaze Game in Shape," *Washington Post*, July 17, 2015, https://www.washingtonpost.com/blogs/erik-wemple/wp/2015/07/17/conde-nast-exec-story-gawker-is-keeping-its-sleaze-game-in-shape/.

49. Gabriel Arana, "Gawker's Outing of Condé Nast's CFO Is Gay-Shaming, Not Journalism," *Huffington Post*, July 18, 2015, http://www.huffingtonpost.com.au/entry/gawker-conde-nast-david-geithner_us_55a90c56e4b0c5f0322d0b2c. See also: Edson C Tandoc and Joy Jenkins, "Out of Bounds? How Gawker's Outing a Married Man Fits into the Boundaries of Journalism," *New Media & Society* 20, no. 2 (2018): 581–598.

50. Remy Stern, "I Had a One-Night Stand with Christine O'Donnell," *Gawker*, October 28, 2010, http://gawker.com/5674353/i-had-a-one-night-stand-with-christine-odonnell.

51. Richard Adams, "Gawker's Christine O'Donnell Sex Smear Marks a New Low" *Guardian* (blog), October 30, 2010, https://www.theguardian.com/world/richard-adams-blog/2010/oct/29/christine-odonnell-gawker-one-night-stand-dustin-dominiak.

52. J. K. Trotter, "Univision Executives Vote to Delete Six Gawker Media Posts," *Gizmodo*, September 10, 2016, http://gizmodo.com/univision-executives-vote-to-delete-six-gawker-media-po-1786466510.

53. http://gizmodo.com/5887480/the-inventor-of-email-did-not-invent-email (article no longer available due to pending litigation).

54. Mario Aguilar, "Internet Pioneers Slam $750,000 Settlement for the 'Man Who Invented Email,'" *Gizmodo*, November 4, 2016, http://gizmodo.com/internet-pioneers-slam-750-000-settlement-for-the-man-1788503950.

55. Jonathan Coopersmith, to SIGCIS members, November 6, 2016, "Gawker Removing History," http://lists.sigcis.org/pipermail/members-sigcis.org/2016-November/000364.html.

56. United States District Court District of Massachusetts. Shiva Ayyadurai, Plaintiff v. Floor64, Inc. d/b/a Techdirt, Michael David Masnick, Leigh Beadon, and DOES 1–20 [reference to dates is on page 4], Civil Action No. 17–10011-FDS.

57. Ayyadurai v. Floor64, Inc., et al., 39–45.

58. Antonioua and Akrivosb, "Hulk Hogan and the Demise of Gawker Media," 153.

59. Joe Mullin and Cyrus Farivar, "History by Lawsuit: After Gawker's Demise, the "Inventor of E-mail" Targets Techdirt," *Ars Technica*, June 13, 2017, https://arstechnica.com/tech-policy/2017/06/shivas-war-one-mans-quest-to-convince-the-world-that-he-invented-e-mail/.

60. United States District Court District of Massachusetts. SHIVA AYYADURAI Plaintiff, 16.

61. Cyrus Farivar, "Judge Dismisses Shiva 'I Invented EMAIL' Ayyadurai's Libel Lawsuit against Techdirt," *Ars Technica*, September, 7, 2017, https://arstechnica.com/tech-policy/2017/09/judge-dismisses-libel-lawsuit-filed-by-self-proclaimed-e-mail-inventor/.

62. Mike Masnick, "The Latest on Shiva Ayyadurai's Failed Libel Suit against Techdirt," *Techdirt*, October 6, 2017, https://www.techdirt.com/articles/20171006/11584638359/latest-shiva-ayyadurais-failed-libel-suit-against-techdirt.shtml.

63. Matthew D. Bunker, Paul H. Gates Jr., and William C. Nevin, "Anti-SLAPP Statutes Offer Tool for Media Defendants," *Newspaper Research Journal* 35 no. 1 (2014): 6–19.

64. Benjamin Mullin, "Bustle Owner Plans to Double Down on Gawker," *Wall Street Journal*, September 17, 2018, https://www.wsj.com/articles/bustle-owner-plans-to-double-down-on-gawker-1537211032.

65. Janet Abbate, email to author, December 15, 2016.

66. See, for example, respectively, Mar Hicks, *Programmed Inequality: How Britain Discarded Women Technologists and Lost Its Edge in Computing* (Cambridge, MA: MIT Press, 2017); Ignaz Strebel, Alain Bovet, and Philippe Sormani, *Repair Work Ethnographies Revisiting Breakdown, Relocating Materiality* (London: Palgrave Macmillan, 2019); Jussi Parikka, *What Is Media Archaeology?* (Cambridge: Polity, 2012); Ahmed Ragab, "Islam Intensified: Snapshot Historiography and the Making of Muslim Identities," *Postcolonial Studies* 22, no. 2 (2019): 203–219; Richard A Grusin, ed., *The Nonhuman Turn* (Minneapolis: University of Minnesota Press, 2015).

67. Hayden White, "The Question of Narrative in Contemporary Historical Theory," *History and Theory* 23, no. 1 (1984): 1–33.

68. Merav Katz-Kimchi, "'Singing the Strong Light Works of [American] Engineers': Popular Histories of the Internet as Mythopoetic Literature," *Information & Culture* 50, no. 2 (2015): 160–180 (special issue on 'Histories of the Internet edited by Thomas Haigh, Andrew L. Russell, and William H. Dutton).

69. V.A. Shiva Ayyadurai, Twitter post, March 7, 2016, 10:32 a.m., https://twitter.com/va_shiva/status/706623446340517888.

70. Laura Crimaldi, "Entrepreneur Shiva Ayyadurai Will Run for Senate as an Independent," *Boston Globe*, November 11, 2017, https://www.bostonglobe.com/metro/2017/11/11/entrepreneur-shiva-ayyadurai-announces-senate-run-independent/sBBj2VhbrLkth9ay5mtJsL/story.html.

71. Associated Press, "Elizabeth Warren Defends Releasing DNA Test to Show Native American heritage," *Guardian*, October 20, 2018, https://www.theguardian.com/us-news/2018/oct/20/elizabeth-warren-dna-native-american-massachusetts.

Chapter 3: The Email Industry

1. José van Dijck, *The Culture of Connectivity: A Critical History of Social Media* (Oxford: Oxford University Press, 2013).

2. Unless otherwise stated all such figures cited are in US dollars.

3. The Radicati Group, "Cloud Business Email Market, 2020–2024 Executive Summary," March 2, 2020, http://www.radicati.com/?p=8801.

4. The Radicati Group, "Cloud Business Email Market, 2018–2022 Executive Summary," June 13, 2018, https://www.radicati.com/?p=15440; and "Cloud Business Email Market, 2020–2024."

5. "Cloud Email and Collaboration Revenue Worldwide from 2015 to 2024," https://www.statista.com/statistics/497864/cloud-business-email-market/.

6. Media Release, Transparency Marketing, "Email Marketing Market: Digital Marketing Era to Provide Impetus to Email Marketing Industry," June 2019, https://www.transparencymarketresearch.com/pressrelease/email-marketing-industry.htm. All figures quoted in this section are in US dollars.

7. Alex Konrad, "The Cloud 100," *Forbes*, September 11, 2019, https://www.forbes.com/cloud100/#7b944d5b5f94.

8. Alexa, "Google.com" Analysis, accessed September 30 2019, https://www.alexa.com/siteinfo/google.com.

9. Gmail Twitter Post, October 27, 2018, https://twitter.com/gmail/status/1055806807174725633.

10. Statista, "Number of Active Gmail Users Worldwide from January 2012 to October 2018," January 18, 2019, https://www.statista.com/statistics/432390/active-gmail-users/.

11. MarketLine, "Alphabet Inc: Company Profile," October 30, 2018, accessed January 10, 2019, from MarketLine Advantage database, p. 3.

12. MarketLine, "Alphabet Inc: Company Profile," 32.

13. MarketLine, "Company Profiles: Technology and Services," accessed January 11, 2019, from MarketLine Advantage database.

14. Litmus, "2018 State of Email Service Providers," accessed January 11, 2019, p. 15.

15. Jason Feifer, "How 'Structure" Saved Mailchimp," *Entrepreneur* 46, no. 6 (2018): 31.

16. Jake Chessum, "Want to Prove Patience Pays Off? Ask the Founders of This 17-Year-Old $525 Million Email Empire!" *Inc.*, Winter 2017, https://www.inc.com/magazine/201802/mailchimp-company-of-the-year-2017.html.

17. MarketLine, "Constant Contact, Inc: Company Profile," December 12, 2018, accessed January 14, 2019, from MarketLine Advantage database, p. 5.

18. Jeff Bauter Engel, "Constant Contact Cuts 15 Percent of Staff after Its Acquisition," *Xconomy* February 12 2016, https://xconomy.com/boston/2016/02/12/constant-contact-cuts-15-percent-of-staff-after-its-acquisition/.

19. Litmus, "2018 State of Email Service Providers," accessed January 11, 2019, pp. 5–7.

20. David H. Crocker et al, "Standard for the Format of ARPA Network Text Messages," RFC 733, https://doi.org/10.17487/RFC0733, November 1977.

21. Thomas Haigh, "Protocols for Profit: Web and E-mail Technologies as Product and Infrastructure," in *The Internet and American Business*, ed. William Aspray and Paul E. Ceruzzi (Cambridge, MA: MIT Press, 2008), 118.

22. Haigh, "Protocols for Profit," 119.

23. Bradley Shimmin, *Effective E-Mail Clearly Explained: File Transfer, Security, and Interoperability* (Open Library: AP Professional, 1997).

24. Steve Whittaker, Victoria Bellotti, and Paul Moody, "Introduction to This Special Issue on Revisiting and Reinventing E-Mail," *Human Computer Interaction* 20 (2005): 1–9.

25. Ed Bott, "What Does the New Gmail API Mean for Internet Standards?," *ZDNet*, June 26, 2014, http://www.zdnet.com/article/what-does-the-new-gmail-api-mean-for-internet-standards/.

26. "About Fastmail," https://www.fastmail.com/about/.

27. Bron Gondwana, "JMAP is on the Home Straight," *Fastmail* (blog), December 27, 2018, https://fastmail.blog/2018/12/27/jmap-is-on-the-home-straight/.

28. Write the Docs Conference Australia 2018, "About the Conference," November 15–16, 2018, http://www.writethedocs.org/conf/australia/2018/about/.

29. "Opera Acquires E-mail Service Fastmail.fm," *Hacker News*, May 1, 2010, https://news.ycombinator.com/item?id=1307649.

30. Rob Mueller, "Exciting News: FastMail Staff Purchase the Business from Opera," *Fastmail* (blog), September 25, 2013, https://fastmail.blog/2013/09/25/exciting-news-fastmail-staff-purchase-the-business-from-opera/.

31. Chris Duckett, "FastMail Staff Buy Back Company from Opera," *ZDNet*, September 26, 2013, http://www.zdnet.com/article/fastmail-staff-buy-back-company-from-opera/.

32. Michael Cyger, "After 15 Years, FastMail Finally Acquires Their .Com—With Rod Mueller," *Domain Sherpa*, November 17, 2014, http://www.domainsherpa.com/rob-mueller-fastmail-interview/.

33. Bettina Specht, "The State of Email Engagement," July 8, 2019, https://litmus.com/blog/infographic-the-2019-email-client-market-share.

34. Lauren Smith, "Email Client Market Share: Where People Opened in 2013," *Litmus* (blog), January 16, 2014, https://litmus.com/blog/email-client-market-share-where-people-opened-in-2013.

35. Justine Jordan, "Top 10 Most Popular Email Clients of 2015," *Litmus* (blog), December 15, 2015, https://litmus.com/blog/top-10-most-popular-email-clients-of-2015.

36. "Email Client Market Share," Litmus Email Analytics, http://emailclientmarketshare.com/.

37. Bettina Specht, "The 2019 Email Client Market Share," July 8, 2019, https://litmus.com/blog/infographic-the-2019-email-client-market-share.

38. Watson Marketing, "2018 Marketing Benchmark Report," IBM Corporation, p. 25, https://www.ibm.com/downloads/cas/L2VNQYQ0.

39. "The Ultimate Mobile Email Statistics Overview," *emailmonday*, accessed April 9, 2017, http://www.emailmonday.com/mobile-email-usage-statistics.

40. John Watton, "Adobe Email Survey 2016," *Digital Marketing Blog Europe*, https://blogs.adobe.com/digitaleurope/cross-channel-marketing/adobe-email-survey-2016/.

41. Gavin Lucas, *The Story of Emoji* (Munich: Prestel Publishing, 2016), 10. For a rich historical analysis of visual modes of digital media communication see: Larissa Hjorth, *Mobile Media in the Asia-Pacific: Gender and the Art of Being Mobile* (London: Routledge, 2011).

42. Campaign Monitory, "Using Emoji and Symbols in Your Email Subject Lines," December 2018, https://www.campaignmonitor.com/resources/guides/using-emojis-and-symbols-in-email-marketing/.

43. Katie Hafner and Matthew Lyon, *Where Wizards Stay Up Late* (New York: Simon & Schuster, 1996), 217.

44. Lucas, *The Story Of Emoji*, 25–37.

45. "Word of the Year 2015," https://languages.oup.com/word-of-the-year/word-of-the-year-2015.

46. "Diversify My Emoji," DoSomething.org, accessed April 9, 2017, https://www.dosomething.org/us/campaigns/diversify-my-emoji.

47. "Unicode Emoji, Technical Report #51," *Unicode*, November 22, 2016, http://www.unicode.org/reports/tr51/.

48. Chance Miller, "Facebook Messenger Puts 1500 New, More Diverse Emojis in Its Users' Arsenal," 9to5mac.com, https://9to5mac.com/2016/06/01/diverse-facebook-messenger-emoji/.

49. For example see: Selina Jeanne Sutton, "Emoji are Becoming More Inclusive, but Not Necessarily More Representative," *The Conversation*, February 8, 2019, https://theconversation.com/emoji-are-becoming-more-inclusive-but-not-necessarily-more-representative-111388; Andrew McGill, "Why White People Don't Use White Emoji," *The Atlantic*, May 9, 2016, https://www.theatlantic.com/politics/archive/2016/05/white-people-dont-use-white-emoji/481695/; Paige Tutt, "Apple's New Diverse Emoji Are Even More Problematic Than Before," *Washington Post*, April 10, 2015, https://www.washingtonpost.com/posteverything/wp/2015/04/10/how-apples-new-multicultural-emojis-are-more-racist-than-before/.

50. Ad Council, "I Am a Witness Campaign," accessed April 9, 2017, http://iwitnessbullying.org/about.

51. Luke Stark and Kate Crawford, "The Conservatism of Emoji: Work, Affect, and Communication," *Social Media + Society* 1, no. 2 (2015): 8.

52. Umashanthi Pavalanathan and Jacob Eisenstein, "More Emojis, less :) The Competition for Paralinguistic Function in Microblog Writing," *First Monday* 21, no. 11 (2016).

53. Petra Kralj Novak, Jasmina Smailović, Borut Sluban, and Igor Mozetič, "Sentiment of Emojis," *PLOS ONE* 10, no. 12 (2015): 1–22.

54. "Mailchimp's Most Popular Subject Line Emojis," *Mailchimp* (blog), May 4, 2015, accessed April 10, 2017, https://mailchimp.com/resources/mailchimps-most-popular-subject-line-emojis/.

55. A number of scholars, however, have critiqued emoji universalization. See: Tim Highfield and Tama Leaver, "Instagrammatics and Digital Methods: Studying Visual Social Media, from Selfies and GIFs to Memes and Emoji," *Communication Research and Practice* 2, no. 1 (2016): 47–62.

56. Kayla Lewkowicz, "Harness the Power of Emojis in Your Inbox," *Litmus* (blog), February 13, 2017, https://litmus.com/blog/harness-the-power-of-emojis-in-your-inbox.

57. "How We Set Up Emoji Support for Subject Lines," *Mailchimp* (blog), February 25, 2015, https://mailchimp.com/resources/how-we-set-up-emoji-support-for-subject-lines/.

58. Mailchimp, "How We Set Up Emoji Support for Subject Lines."

59. Terje Rasmussen, *Personal Media and Everyday Life: A Networked Lifeworld* (London: Palgrave Macmillan, 2014), 32.

60. "comScore Releases First Comparative Report On Mobile Usage in Japan, United States and Europe," comScore.com, http://www.comscore.com/Insights/Press-Releases/2010/10/comScore-Release-First-Comparative-Report-on-Mobile-Usage-in-Japan-United-States-and-Europe.

61. Kenichi Ishii, "Implications of Mobility: The Uses of Personal Communication Media in Everyday Life," *Journal of Communication* 56 (2006): 346–365.

62. Mizuko Ito and Daisuke Okabe, "Technosocial Situations: Emergent Structuring of Mobile E-mail Use," in *Personal, Portable Intimate: Mobile Phones in Japanese Life*, ed. Mizuko Ito, Misa Matsuda, and Daisuke Okabe (Cambridge, MA: MIT Press, 2005).

63. Lars A. Knutsen and Kalle Lyytinen, "Messaging Specifications, Properties and Gratifications as Institutions: How Messaging Institutions Shaped Wireless Service Diffusion in Norway and Japan," *Information and Organization* 18 (2008): 116.

64. Kakuko Miyata, Jeffrey Boase, and Barry Wellman, "The Social Effects of Keitai and Personal Computer E-mail in Japan," in *Handbook of Mobile Communication Studies*, ed. James E. Katz (Cambridge MA: MIT Press, 2008), 209–223. Although published in 2008, draft chapters of the research were circulated in 2006.

65. Gerard Goggin and Larissa Hjorth, "Introduction: Mobile Media Research—State of the Art," in *The Routledge Companion to Mobile Media*, ed. Gerard Goggin and Larissa Hjorth (New York: Routledge, 2013), 1–8.

66. Kyoung-hwa Yonnie Kim, "Genealogy of Mobile Creativity: A Media Archaeological Approach to Literary Practice in Japan," in *The Routledge Companion to Mobile Media*, ed. Gerard Goggin and Larissa Hjorth (New York: Routledge, 2013), 216–224.

67. Matt Alt, *The Secret Lives of Emoji: How Emoticons Conquered the World*, Amazon Digital Services, Kindle Edition, 2016, "Part Three: Emoji Get Their Greencard."

68. "How Are Mobile, Desktop and Webmail Categories Defined?," *Litmus*, https://litmus.com/help/analytics/category-definitions/.

69. "How Gmail's New Inbox is Affecting Open Rates," *Mailchimp* (blog), July 23, 2013, https://web.archive.org/web/20130828150415/http://blog.mailchimp.com/how-gmails-new-inbox-is-affecting-open-rates/.

70. "About Gmail Tabs," *Mailchimp*, accessed March 2, 2017, https://mailchimp.com/help/about-gmail-tabs.

71. "Which Gmail Tab Will Your Email Appear Under?," *Litmus*, https://litmus.com/gmail-tabs.

72. Eric Ravenscraft, "Disable Automatic Image Loading In Gmail To Improve Security," *Lifehacker*, December 15, 2013, http://lifehacker.com/disable-automatic-image-loading-in-gmail-to-save-data-a-1482522063.

Chapter 4: Bureaucratic Intensity and Email in the Workplace

1. Scott Rosenberg, "Shut Down Your Office. You Now Work in Slack," *Backchannel*, May 8, 2015, https://backchannel.com/shut-down-your-office-you-now-work-in-slack-fa83cb7cce6c#.uwm9wd887.

2. Norman Z. Shapiro and Robert H. Anderson, *Toward an Ethics and Etiquette for Electronic Mail* (Santa Monica, CA: RAND, 1985), 9–10.

3. Lee Sproull and Sara Kiesler, "Reducing Social Context Cues: Electronic Mail in Organizational Communication," *Management Science* 32, no. 11 (1986): 1492–1512.

4. While I am aware of the apparent problems of using the terms "affect" and "emotion" interchangeably, and of the substantial critical literature that supports maintaining these distinctions, I am not convinced. For me, the dismissal of emotion (for that is always the outcome—emotion is never positioned as superior or even simply equal in the taxonomies of affect) is part of a critical enterprise whose mission has now passed. More significantly it speaks to a continued eclipse of the role of caring and feeling within organizational life. For compelling discussions of this dichotomy, see: Ruth Leys, "The Turn to Affect: A Critique," *Critical Inquiry* 37, no. 3 (2011): 434–472; and Sara Ahmed, "Open Forum Imaginary Prohibitions: Some Preliminary Remarks on the Founding Gestures of the 'New Materialism,'" *European Journal of Women's Studies* 15, no. 1 (2008): 23–39.

5. Lisa Gitelman, *Paper Knowledge: Toward a Media History of Documents* (Durham, NC: Duke University Press, 2014).

6. Merje Kuus, "Transnational Bureaucracies: How Do We Know What They Know?" *Progress in Human Geography* 39, no. 4 (2015): 432–448.

7. Ben Kafka, *The Demon of Writing: Powers and Failures of Paperwork* (Brooklyn, NY: Zone Books, 2012).

8. Michael E. Holmes, "Don't Blink or You'll Miss It: Issues in Electronic Mail Research," *Communication Yearbook* 18, no. 1 (1995): 454–463. See also: Ronald Rice, "Computer-Mediated Communication and Organizational Innovation," *Journal of Communication* 37, no. 4 (Autumn 1987): 65–94.

9. Laura Garton and Barry Wellman, "Social Impacts of Electronic Mail in Organisations: A Review of the Research Literature," *Communication Yearbook* 18 (1995): 434–453.

10. John Sherblom, "Direction, Function, and Signature in Electronic Mail," *International Journal of Business Communication* 25, no. 4, (1988): 39–54.

11. Joan Waldvogel, "Greetings and Closings in Workplace Email," *Journal of Computer Mediated Communication* 12, no. 2 (January 2007): 456–477.

12. Nava Pliskin, Celia T Romm, and Raymond Markey, "E-mail as a Weapon in an Industrial Dispute," *New Technology, Work and Employment* 12, no. 1 (1997).

13. Eric Gilbert, "Phrases That Signal Workplace Hierarchy," Proceedings of the ACM 2012 Conference on Computer Supported Cooperative Work, New York (2012): 1037–1046.

14. Nicolas B. Ducheneaut, "The Social Impacts of Electronic Mail in Organizations: A Case Study of Electronic Power Games Using Communication Genres," *Information, Communication & Society* 5, no. 2 (2002): 153–188.

15. Starr Roxanne Hiltz and Murray Turoff, "Structuring Computer-Mediated Communication Systems to Avoid Information Overload," *Communications of the ACM* 28, no. 7 (1985): 680–689; Stephen R. Barley, Debra E. Meyerson, and Stine Grodal, "E-Mail as a Source and Symbol of Stress," *Organization Science* 22, no. 4 (2011): 887–906; Jean-François Stich, Monideepa Tarafdar, Cary L Cooper, and Patrick Stacey, "Workplace Stress from Actual and Desired Computer-Mediated Communication Use: A Multi-Method Study," *New Technology, Work & Employment* 32, no. 1 (2017): 84–100; Rita S. Mano and Gustavo Mesch, "E-mail characteristics, work performance and distress," *Computers in Human Behavior* 26 (2010): 61–69.

16. Judy Wajcman, *Pressed for Time: The Acceleration of Life in Digital Capitalism* (Chicago: University of Chicago Press: 2015), 98.

17. Melissa Gregg, *Counterproductive: Time Management in the Knowledge Economy* (Durham, NC: Duke University Press, 2018), 62.

18. Polina Zilberman, Gilad Katz, Asaf Shabtai, and Yuval Elovici, "Analyzing Group E-mail Exchange to Detect Data Leakage," *Journal of the American Society for Information Science and Technology* 64, no. 9 (2013): 1780–1790; Louise L. Hill, "Personal Use of Workplace Computers: A Threat to Otherwise Privileged Communications," *Journal of Internet Law* 15, no. 9 (2012): 20–31.

19. Melissa A, Mazmanian, Wanda J. Orlikowski, and JoAnne Yates, "Crackberries: The Social Implications of Ubiquitous Wireless e-Mail Devices," in *Designing Ubiquitous Information Environments: Socio-Technical Issues and Challenges*, ed. C. Sørensen, Y. Yoo, K. Lyytinen, and J. DeGross (New York: Springer, 2005), 337–343.

20. Melissa Mazmanian, Wanda Orlikowski, and Joanne Yates, "The Autonomy Paradox: The Implications of Mobile Email Devices for Knowledge Professionals," *Organization Science* 24, no. 5 (2013): 1337–1357. See also: Catherine A Middleton and Wendy Cukieris, "Is Mobile Email Functional or Dysfunctional? Two Perspectives on Mobile Email Usage," *European Journal of Information Systems* 15, no. 3 (2006): 252–260; and Jeff Funtasz, "Canadian Middle Manager Experience with Mobile Email Technologies," *Information, Communication & Society* 15, no. 8 (2012): 1217–1235.

21. Martin Svensson and Alf Westelius, "@ the Emotional Verge: When Enough Is Enough in Email Conversations," in *Individual Sources, Dynamics and Expressions of Emotions. Research on Emotion in Organizations*, ed. E. J. Charmine Härtel, Neal M. Ashkanasy, and W. J. Zerbe (Bingley, UK: Emerald Group Publishing Limited, 2013); Karianne Skovholt, Anette Grønning, and Anne Kankaanranta, "The Communicative Functions of Emoticons in Workplace E-Mails: :-)," *Journal of Computer-Mediated Communication* 19, no. 4 (2014): 780–797; V. K. G. Lim and T. S. H, Teo, "Mind Your

E-manners: Impact of Email Incivility on Employees' Work Attitude and Behavior," *Information Management* 46 (2009): 419–425.

22. YoungAh Park and Verena C. Haun, "The Long Arm of Email Incivility: Transmitted Stress to the Partner and Partner Work Withdrawal," *Journal of Organizational Behavior* 39, no. 10 (2018): 1268–1282.

23. S. Shirren and James G. Phillips "Decisional Style, Mood and Work Communication: Email Diaries," *Ergonomics* 54, no. 10 (2011): 891–903, https://doi.org/10.1080/00140139.2011.609283.

24. "Intermedia's 2018 Workplace Communications Report: How Technology is Redefining the Workplace, Workday, and Workforce," https://www.intermedia.net/report/ucaasworkplace2018.

25. "The Way We Work: 9,000 Knowledge Workers Share Their Insight on the Jobs They Do and the Workplace of the Future," *Unify White Paper*, accessed May 5, 2017, http://www.economyup.it/upload/images/11_2016/161122122232.pdf.

26. Unify, "The Way We Work 2018 Topics Worthy of Discussion Series. Guest blog with Tim Banting," May 28, 2018, https://unify.com/en/blog/expert-analyst-shares-unique-insights-on-the-way-we-work-in-2018-and-beyond.

27. "Okta's Businesses @ Work Report," Okta, accessed September 14, 2019, https://www.okta.com/businesses-at-work/2019/.

28. Public Relations Society of America, *2017 Report, Technology Trends in the Communication Industry*, https://www.theemployeeapp.com/wp-content/uploads/2017/08/Technology-Trends-in-the-Communications-Industry-Report_Final-.pdf.

29. Media Release, GFI Software, "Work Email Onslaught: Staff Have Nowhere to Hide, US Study Finds," June 24, 2015, https://www.gfi.com/company/press/press-releases/2015/06/work-email-onslaught-staff-have-nowhere-to-hide-us-study-finds.

30. "Americans Stay Connected to Work on Weekends, Vacation and Even When Out Sick," American Psychological Association, September 4, 2013, http://www.apa.org/news/press/releases/2013/09/connected-work.aspx.

31. "Is Work Email Disrupting the Personal Lives of US Employees?," *Samanage*, March 30, 2016, https://onlabor.org/wp-content/uploads/2017/03/Samanage-Email-Overload-Survey.pdf.

32. Samanage, ""Employees Squander More Than a Month Each Year Checking Email Outside of Work Hours," *PR Newswire*, April 21, 2016, https://www.prnewswire.com/news-releases/samanage-survey-shows-employees-squander-more-than-a-month-each-year-checking-email-outside-of-work-hours-300255026.html.

33. Liuba Y. Belkin, William J. Becker, and Samantha A. Conroy, "Exhausted, But Unable to Disconnect: After-Hours Email, Work-Family Balance and Identification,"

Academy of Management (Proceedings, 2016), https://doi.org/10.5465/AMBPP.2016.10353abstract.

34. Keri K. Stephens, *Negotiating Control: Organizations and Mobile Communication* (New York: Oxford University Press, 2018), 219.

35. Emily van der Nagel, "'Networks That Work Too Well': Intervening in Algorithmic Connections," *Media International Australia* 168, no. 1 (2018): 81–92.

36. Matthew Fuller and Andrew Goffey, *Evil Media* (Cambridge, MA: MIT Press, 2012), 15.

37. Vincent de Gournay, conversion quoted with Melchior von Grimm, in Kafka, *The Demon of Writing,* 77.

38. Kafka, *The Demon of Writing,* 79.

39. Paul du Gay, *In Praise of Bureaucracy: Weber, Organization, Ethics* (London: Sage, 2000), 1–13.

40. Du Gay, *In Praise of Bureaucracy,* 64–65.

41. Du Gay, 66–68.

42. Antonino Palumbo and Alan Scott, "Bureaucracy, Open Access and Social Pluralism: Returning the Common to the Goose," in *The Values of Bureaucracy,* ed. Paul Du Gay (Oxford: Oxford University Press, 2005), 281–308.

43. Charles T. Goodsell, *The New Case for Bureaucracy* (Los Angeles: Sage/CQ Press, 2015).

44. Michael Herzfeld, *The Social Production of Indifference: Exploring the Symbolic Roots of Western Bureaucracy* (Chicago: University of Chicago Press, 1992), 5.

45. Ralph P. Hummel, *The Bureaucratic Experience: The Post-modern Challenge* (Armonk, NY: M. E. Sharpe, 2008), 18.

46. Janet Newman, "Bending Bureaucracy: Leadership and Multi-Level Governance," in *The Values of Bureaucracy,* ed. Paul Du Gay (Oxford: Oxford University Press, 2005), 191–210.

47. Catherine J. Turco, *The Conversational Firm: Rethinking Bureaucracy in the Age of Social Media* (New York: Columbia University Press, 2016), 8.

48. Turco, *The Conversational Firm,* 2.

49. Turco, 187.

50. Turco, 209.

51. Turco, 210–211.

52. Turco, 208.

53. Turco, 8–9.

54. Turco, 128.

55. Turco, 133.

56. MarketLine, "HubSpot Inc: Company Profile," April 6, 2018, accessed January 26, 2019 from MarketLine Advantage database, 3–20.

57. David Z. Morris, "FBI Documents Detail HubSpot's Alleged Attacks on 'Fake Steve Jobs' Writer," *Fortune*, March 26, 2016, http://fortune.com/2016/03/26/fbi-hubspot-fake-steve-jobs/.

58. Du Gay, *In Praise of Bureaucracy*, 3.

59. Du Gay, 11.

60. For example see: Brooke Erin Duffy and Emily Hund, "'Having it All' on Social Media: Entrepreneurial Femininity and Self-Branding Among Fashion Bloggers," *Social Media + Society* 1, no. 2 (2015): 1–11; Melissa Gregg, "The Normalisation of Flexible Female Labour in the Information Economy," *Feminist Media Studies* 8, no. 3, (2008): 285–299; and Angela McRobbie, *The Aftermath of Feminism* (London: Sage, 2008).

61. For example see: Kathi Weeks, "Life Within and Against Work: Affective Labor, Feminist Critique, and Post-Fordist Politics," *ephemera* 7, no. 1 (2007): 233–249.

62. Sophie Bourgault, "Prolegomena to a Caring Bureaucracy," *European Journal of Women's Studies* 24, no. 3 (2016): 207–212, https://doi.org/10.1177/1350506816643730; and Joan C Tronto, *Caring Democracy Markets, Equality, and Justice* (New York: NYU Press, 2013).

63. Camilla Stivers, *Gender Images in Public Administration: Legitimacy and the Administrative State*, 2nd ed. (Thousand Oaks, CA: Sage, 2002), 58.

64. Kathy Ferguson, *The Feminist Case Against Bureaucracy* (Philadelphia: Temple University Press, 1984), 108–109.

65. Mike Savage and Anne Witz, eds., *Gender and Bureaucracy* (London: Blackwell, 1992).

66. Noelie Maria Rodriguez, "Transcending Bureaucracy: Feminist Politics at a Shelter for Battered Women," *Gender & Society* 2, no. 2 (1988): 214–227; Yvonne Due Billing, "Gender Equity—A Bureaucratic Enterprise?," in *The Values of Bureaucracy*, ed. Paul Du Gay (Oxford: Oxford University Press, 2005); Karen Lee Ashcraft, "Organized Dissonance: Feminist Bureaucracy as Hybrid Form," *Academy of Management* 44, no. 6 (2001): 1301–1322.

67. Anthony McCosker, *Intensive Media: Aversive Affect and Visual Culture* (Basingstoke: Palgrave Macmillan, 2013), 20.

68. Tekla Perry, "E-mail Pervasive and Persuasive." *IEEE Spectrum* 29, no. 10 (October 1992): 22–23.

69. Michael Y. Lee and Amy C. Edmondson, "Self-Managing Organizations: Exploring the Limits of Less-Hierarchical Organizing, *Research in Organizational Behavior* 37 (2017): 35–58.

Chapter 5: Moderation and Governance in Email Discussion Forums

1. Alan Sondheim, to Cybermind members, January 2, 2017, "Note to Jon Marshall about Living on Cybermind," https://listserv.wvu.edu/cgi-bin/wa?A0=CYBERMIND.

2. Jonathan Paul Marshall, *Living on Cybermind: Categories, Communication, and Control* (New York: Peter Lang, 2007).

3. Esther Milne, *Letters, Postcards, Email: Technologies of Presence* (London: Routledge, 2010).

4. Ella Taylor-Smith and Colin Smith, "Non-public eParticipation in Social Media Spaces," in *SMSociety '16: Proceedings of the 7th 2016 International Conference on Social Media & Society*, no. 3 (July 2016): 1–8, https://doi.org/10.1145/2930971.2930974.

5. The Association of Internet Researchers, Air-L Archives, http://listserv.aoir.org/pipermail/air-l-aoir.org/.

6. Wikimedia Foundation Mailing List, The Wikimedia-l Archives, https://lists.wikimedia.org/pipermail/wikimedia-l/.

7. The Internet Engineering Task Force, https://www.ietf.org/list/.

8. VICTORIA 19th-Century British Culture & Society, Indiana University Mailing List, https://list.indiana.edu/sympa/arc/victoria.

9. Gmane, http://www.gmane.org/; MARC, http://marc.info/?q=about#What; The Mail Archive, https://www.mail-archive.com/.

10. Darya Gudkova, "Kaspersky Security Bulletin. Spam Evolution 2013," *Securelist*, January 23, 2104, https://securelist.com/58274/kaspersky-security-bulletin-spam-evolution-2013/.

11. Igor Helman, "Spam-A-Lot: The States' Crusade Against Unsolicited Email in Light of the CAN-SPAM Act and the Overbreadth Doctrine," *Boston College Law Review* 50, no. 5 (2009): 1525, n1.

12. "The Definition of Spam," Spamhaus, https://www.spamhaus.org/consumer/definition/.

13. Statista, "Global Spam Volume as Percentage of Total E-mail Traffic from 2007 to 2018," https://www.statista.com/statistics/420400/spam-email-traffic-share-annual/.

14. "Common Rookie Mistakes for Email Marketers," *Mailchimp*, accessed April 13, 2020, https://mailchimp.com/resources/common-rookie-mistakes-email-marketers/.

15. Finn Brunton, *Spam: A Shadow History of the Internet* (Cambridge, MA: MIT Press, 2013), 31.

16. Brunton, *Spam*, 199.

17. Brunton, 70.

18. Igor Helman, "Spam-A-Lot: The States' Crusade Against Unsolicited Email in Light of the CAN-SPAM Act and the Overbreadth Doctrine," *Boston College Law Review* 50, no. 5 (2009): 1537–1538, n1.

19. Brunton, *Spam*, 99–100.

20. Alex C. Kigerl, "Deterring Spammers: Impact Assessment of the CAN SPAM Act on Email Spam Rates," *Criminal Justice Policy Review* 27, no. 8 (2016): 791–811.

21. Lisa R. Lifshitz, "CASL—How To Send E-mails to Canadians Safely," *Business Law Today*, April 2016, http://www.americanbar.org/publications/blt/2016/04/04_lifshitz.html.

22. Theresa Bugeaud and Jonathan Benton, "Benchmarks: The 2018 Report of How Nonprofits are Performing Online," M&R and Nonprofit Technology Network (2018).

23. Brett Schenker, "The 2017 Nonprofit Email Deliverability Study: How Much Does Spam Hurt Online Fundraising?" EveryAction, (2017).

24. Cristina Faba-Pérez and Ana-María Cordero-González, "The Validity of Bradford's Law in Academic Electronic Mailing Lists," *The Electronic Library* 33 no. 6 (2015): 1043–1044.

25. Uwe Matzat, "Quality of Information in Academic E-mailing Lists," *Journal of the American Society for Information Science and Technology* 60, no. 9 (2009): 1859–1870.

26. Uwe Matzat, "Disciplinary Differences in the Use of Internet Discussion Groups: Differential Communication Needs or Trust Problems?" *Journal of Information Science* 35, no. 5 (2009): 613–631.

27. Wei Ding, Peng Liang, Antony Tang, and Hans Van Vliet, "Understanding the Causes of Architecture Changes using OSS Mailing Lists," *International Journal of Software Engineering and Knowledge Engineering* 25, no. 9 (2015): 1633–1651.

28. Nora McDonald, "Distributed Leadership in OSS," *Proceedings of the 18th International Conference on Supporting Group Work* (New York: ACM Digital Library, 2014): 261–262.

29. Emad Shihab, Nicholas Bettenburg, Bram Adams, and Ahmed E. Hassan, "On the Central Role of Mailing Lists in Open Source Projects: An exploratory study," in *New Frontiers in Artificial Intelligence*, ed. Kumiyo Nakakoji, Yohei Murakami and Eric McCready (Berlin: Springer, 2010), 91–103.

30. Anja Guzzi, Alberto Bacchelli, Michele Lanza, Martin Pinzger, and Arie van Deursen, "Communication in Open Source Software Development Mailing Lists," Proceedings of the 10th Working Conference on Mining Software Repositories, San Francisco, California, March 18–19, 2013.

31. Kate Crawford, "Following You: Disciplines of Listening in Social Media," *Continuum* 23, no. 4 (2009): 525–535, https://doi.org/10.1080/10304310903003270.

32. Karl Gutwin, Reagan Penner, and Kevin Schneider, "Group Awareness in Distributed Software Development," Proceedings from ACM Computer-Supported Cooperative Work Conference, Chicago, Illinois, 2004.

33. Amelia Johns and Abbas Rattani, "Somewhere in America": The #MIPSTERZ Digital Community and Muslim Youth Voice Online," in *Negotiating Digital Citizenship: Control, Contest and Culture*, ed. Anthony McCosker, Sonja Vivienne, and Amellia Johns (London: Rowman & Littlefield, 2016).

34. "Yahoo! Groups," *Wikipedia*, last modified May 16, 2017, https://en.wikipedia.org/wiki/Yahoo!_Groups. Unfortunately, despite repeated inquiries, "Yahoo! Research Labs" were unable to provide me with any recent site statistics or usage figures for Yahoo! Groups. Perhaps such reluctance reflects the widespread speculation from subscribers of these groups that Yahoo! intends to kill the service. A number of the moderators expressed frustration because Yahoo! provides no technical support for any of the group management features.

35. Tiziana Terranova, "Free Labor: Producing Culture for the Digital Economy," *Social Text* 18, no. 2 (2000): 33–34.

36. Hector Postigo, "America Online Volunteers: Lessons From an Early Co-production Community," *International Journal of Cultural Studies* 12, no. 5 (2009): 453.

37. Postigo, "American Online Volunteers."

38. Postigo, 455.

39. Postigo, 460–461.

40. Lauren Kirchner, "AOL Settled with Unpaid 'Volunteers' for $15 Million," *Columbia Journalism Review*, February 10, 2011.

41. Brett Neilson and Ned Rossiter, "From Precarity to Precariousness and Back Again: Labour, Life and Unstable Networks," *The Fibreculture Journal* no. 5 (2005). See also: Rosalind Gill and Andy Pratt, "In the Social Factory? Immaterial Labour, Precariousness and Cultural Work," *Theory, Culture & Society* 25, no. 7–8 (2008): 1–30; Ned Rossiter, "Logistical Worlds," *Cultural Studies Review* 20, no. 1 (2014): 53–76; and Maurizio Lazzarato, "Immaterial Labor," in *Radical Thought in Italy: A Potential Politics*, ed. P. Virno and M. Hardt, (Minneapolis: University of Minnesota Press, 1996).

42. Hector Postigo, "America Online Volunteers," *International Journal of Cultural Studies* 12, no. 5 (2009): 452.

43. J. Nathan Matias, "The Civic Labor of Online Moderators," Paper presented at The Internet Politics and Policy Conference, Oxford, 2016.

44. Kylie Jarrett, *Feminism, Labour and Digital Media: The Digital Housewife* (New York: Routledge, 2016), 57–78.

45. Jarrett, *Feminism, Labour, and Digital Media*, 65.

46. Jessamy Gleeson, "(Not) 'Working 9–5': The Consequences of Contemporary Australian-based Online Feminist Campaigns as Digital Labour," *Media International Australia* 161, no. 1 (2016): 83.

47. Zane L. Berge and Mauri P. Collins, "Perceptions of E-moderators about Their Roles and Functions in Moderating Electronic Mailing Lists," *Distance Education* 21, no. 1 (2000): 81–100; Sam Kininmonth, "Benevolent Dictatorships? Moderation as a Tool of Platform Governance on Reddit," (Honours Thesis, Swinburne University, 2016).

48. Yahoo! Groups, "DNA-NEWBIE," The International Society of Genetic Genealogy, accessed May 28, 2017, https://groups.yahoo.com/neo/groups/DNA-NEWBIE/info.

49. ECOLOG-L@LISTSERV.UMD.EDU, accessed May 28, 2017, Ecological Society of America, http://www.lsoft.com/scripts/wl.exe?SL1=ECOLOG-L&H=LISTSERV.UMD.EDU.

50. nettime mailing lists, accessed May 28, 2017, http://www.nettime.org/.

51. Yahoo! Groups, "Crossfire—Tabletop Miniature Wargaming," accessed May 28, 2017, https://groups.yahoo.com/neo/groups/Crossfire-WWII/info.

52. Victoria Research Web, "The Electronic Conference for Victorian Studies," accessed May 28, 2017, http://victorianresearch.org/discussion.html.

53. Nick Couldry, *Media, Society, World: Social Theory and Digital Media Practice* (Cambridge: Polity, 2013), 17.

54. Andrew Russeth, "E-flux Launches E-flux Conversations Site, Seeking Debate Away from Social Media, *ARTnews*, March 9, 2015, http://www.artnews.com/2015/03/09/e-flux-launches-e-flux-conversations-site-seeking-discussion-debate/.

Chapter 6: The Enron Database and Hillary Clinton's Emails

1. Christena E. Nippert-Eng, *Islands of Privacy* (Chicago: University of Chicago Press, 2010), 4–5.

2. Thomas Haigh, "Protocols for Profit: Web and E-mail Technologies as Product and Infrastructure," in *The Internet and American Business*, ed. William Aspray and Paul E. Ceruzzi (Cambridge, MA: MIT Press, 2008), 120.

3. Tejaswini Herath, Rui Chen, Jingguo Wang, Ketan Banjara, Jeff Wilbur, and Raghav Rao, "Security Services as Coping Mechanisms: An Investigation Into User Intention to Adopt an Email Authentication Service," *Information Systems Journal* 24, no. 1 (January 2014): 62.

4. Online Trust Alliance, Internet Society website, https://otalliance.org/system/files/files/resource/documents/ota_glossary2014.pdf.

5. Nick Hopkins, "Deloitte Hit by Cyber-Attack Revealing Clients' Secret Emails," *Guardian*, May 25. 2017, https://www.theguardian.com/business/2017/sep/25/deloitte-hit-by-cyber-attack-revealing-clients-secret-emails.

6. "Sony," WikiLeaks, accessed June 4, 2017, https://wikileaks.org/sony/press/.

7. International Consortium of Investigative Journalists, "Giant Leak of Offshore Financial Records Exposes Global Array of Crime and Corruption," April 3, 2016, https://panamapapers.icij.org/20160403-panama-papers-global-overview.html.

8. Marina Walker Guevara, "ICIJ Releases Database Revealing Thousands of Secret Offshore Companies," The Center for Public Integrity (website), May 9, 2016, https://publicintegrity.org/accountability/icij-releases-database-revealing-thousands-of-secret-offshore-companies/.

9. Andy Greenberg, "How Reporters Pulled Off the Panama Papers, The Biggest Leak in Whistleblower History," *Wired*, April 4, 2016, https://www.wired.com/2016/04/reporters-pulled-off-panama-papers-biggest-leak-whistleblower-history/.

10. WikiLeaks, Twitter post, April 27, 2016, 5:19 a.m., https://twitter.com/wikileaks/status/725298248542490625?lang=en.

11. International Consortium of Investigative Journalists, "One Year on Tax Wars, Follow-Up Investigations and Who Was Actually in the Paradise Papers?," November 5, 2018, https://www.icij.org/investigations/paradise-papers/tax-wars-follow-up-investigations-and-who-was-actually-in-the-paradise-papers/.

12. International Consortium of Investigative Journalists, "Three Years On: Panama Papers Helps Recover More Than $1.2 Billion around the World," April 3, 2019, https://www.icij.org/investigations/panama-papers/panama-papers-helps-recover-more-than-1-2-billion-around-the-world/.

13. See: Special Section, "WikiLeaks: From Popular Culture to Political Economy," *International Journal of Communication* 8 (2014).

14. WikiLeaks, "Leaks," https://wikileaks.org/-Leaks-.html.

15. For example see: "WikiLeaks List: Most Damaging Emails about DNC, Clinton, & Bernie," *Heavy*, July 22, 2016, updated July 24, 2016, http://heavy.com/news/2016/07/wikileaks-emails-clinton-bernie-list-directory-photos-most-damaging-worst-rhode-island-delegate-fec-jvf/.

16. “Enron Fast Facts,” *CNN*, April 27, 2017, http://edition.cnn.com/2013/07/02/us/enron-fast-facts/.

17. “Enron’s Collapse; Excerpts From a House Hearing on Destruction of Enron Documents,” *New York Times*, January 25, 2002, http://www.nytimes.com/2002/01/25/business/25FULL-TEXT-HOUSE.html.

18. “The Fall of Enron,” *NPR*, accessed June 4, 2017, http://www.npr.org/news/specials/enron/.

19. Elisa Moncarz, Raul Moncarz, Alejandra Cabello, and Benjamin Moncarz, “The Rise and Collapse of Enron: Financial Innovation, Errors and Lessons,” *Accounting and Management* (2006): 17–37.

20. Julian Borger, “Tapes Reveal Enron’s Secret Role in California’s Power Blackouts,” *Guardian*, February 5, 2005, http://www.theguardian.com/business/2005/feb/05/enron.usnews.

21. Borger, “Tapes Reveal Enron’s Secret Role.”

22. Anita Raghavan, Kathryn Kranhold, and Alexei Barrionuevo, “How Enron Bosses Created A Culture of Pushing Limits,” *Wall Street Journal*, August 26, 2002, https://www.wsj.com/articles/SB1030320285540885115.

23. Bethany McLean and Peter Elkind, *The Smartest Guys in the Room: The Amazing Rise and Scandalous Fall of Enron* (New York: Portfolio Trade, 2003), 122.

24. Sherron Watkins has been hailed as a whistleblower and was named by *Time* as one of its persons of the year in 2002 for her work revealing the criminal activities of Enron, http://content.time.com/time/specials/packages/0,28757,2022164,00.html. However, other people disagree about the whistleblower status she has been accorded. See: Lynn Brewer and Matthew Scott Hansen, *Confessions of an Enron Executive: A Whistleblower’s Story* (Bloomington, IL: AuthorHouse, 2004).

25. John Alan Cohan, “‘I Didn’t Know’ and ‘I Was Only Doing My Job’: Has Corporate Governance Careened Out of Control? A Case Study of Enron’s Information Myopia,” *Journal of Business Ethics* 40, no. 3 (2002): 275–299.

26. Eric Gilbert, “Phrases That Signal Workplace Hierarchy,” Proceedings of the ACM 2012 conference on Computer Supported Cooperative Work, New York (2012): 1037–1046.

27. Pat Wood, William Massey, Linda Breathitt, and Nora Mead Brownell (Commissioners), “Fact-Finding Investigation of Potential Manipulation of Electric and Natural Gas Prices,” Federal Energy Regulatory Commission, February 13, 2002.

28. Joe Bartling, “The Enron Data Set—Where Did It Come From?” Bartling Forensic LLC, accessed June 7, 2017, http://www.bartlingforensic.com/the-enron-data-set-where-did-it-come-from/.

29. Jessica Leber, "The Immortal Life of the Enron E-mails," *MIT Technology Review*, July 2, 2013, https://www.technologyreview.com/2013/07/02/177506/the-immortal-life-of-the-enron-e-mails/.

30. William W. Cohen, "Enron Email Dataset," Carnegie Mellon University, May 8, 2015, https://www.cs.cmu.edu/~./enron/.

31. Bryan Klimt and Yiming Yang, "The Enron Corpus: A New Dataset for Email Classification Research," 15th European Conference on Machine Learning, Italy, September, 2004, conference proceedings in *Lecture Notes in Computer Science*, vol. 3201 (Springer: Berlin, 2004): 217–226; see also: Ron Bekkerman, Andrew McCallum, and Gary Huang, "Automatic Categorization of Email into Folders: Benchmark Experiments on Enron and SRI Corpora," *Center for Intelligent Information Retrieval*, Technical Report *IR 418* (2004).

32. Jitesh Shetty and Jafar Adibi, "The Enron Email Dataset Database Schema and Brief Statistical Report," *Information Sciences Institute Technical Report* (Los Angeles: University of Southern California, 2004), https://pdfs.semanticscholar.org/0d84/1ba6c3aa123cfcca2a312eab9b9081e6a187.pdf.

33. Will Styler, "The EnronSent Corpus, Technical Report, '01–2011," University of Colorado at Boulder Institute of Cognitive Science, 2011.

34. For example see: Svitlana Volkova, Theresa Wilson, and David Yarowsky, "Exploring Demographic Language Variations to Improve Multilingual Sentiment Analysis in Social Media," Proceedings of the 2013 Conference on Empirical Methods in Natural Language Processing (2013): 1815–1827; Steven J. Castellucci and Scott MacKenzie, "Gathering Text Entry Metrics on Android Devices," Proceedings of the International Conference on Multimedia and Human-Computer Interaction, *MHCI* 120 (2013): 1–8.

35. Yingjie Zhou, "Mining Organizational Emails for Social Networks with Application to Enron Corpus," PhD dissertation, Rensselaer Polytechnic Institute, New York, July 2008.

36. Jana Diesner, Terrill Frantz, and Kathleen Carley, "Communication Networks from the Enron Email Corpus: 'It's Always about the People. Enron is no Different,'" *Computational & Mathematical Organization Theory* 11 (2005): 201–228.

37. Kathleen Carley and David Skillicorn, eds., "Special Issue on Analyzing Large Scale Networks: The Enron Corpus," *Computational & Mathematical Organization Theory* 11 (2005): 179–181.

38. Eric Gilbert, "Phrases That Signal Workplace Hierarchy," Paper presented at CSCW '12, Conference on Computer Supported Cooperative Work, Seattle, Washington, February, 2012.

39. William Deitrick, Zachary Miller, Benjamin Valyou, and Wei Hu, "Author Gender Prediction in an Email Stream Using Neural Networks," *Journal of Intelligent Learning Systems and Applications* 4, no. 3 (2012): 169–175.

40. Michael Berry and Murray Browne, "Email Surveillance Using Non-negative Matrix Factorization," *Computational & Mathematical Organization Theory* 11 (2005): 249–264.

41. Johanna Hardin and Ghassan Sarkis, "Network Analysis with the Enron Email Corpus," *Journal of Statistics Education* 23, no. 2 (2015); Andrew McCallum, Xuerui Wang, and Andres Corrada-Emmanuel, "Topic and Role Discovery in Social Networks with Experiments on Enron and Academic Email," *Journal of Artificial Intelligence Research* 30 (2007): 249–272; Garnett Wilson and Wolfgang Banzhaf, "Discovery of Email Communication Networks from the Enron Corpus with a Genetic Algorithm using Social Network Analysis," Proceedings of the IEEE Congress on Evolutionary Computation Conference, CEC 2009, Trondheim, Norway, May, 2009; Przemysław Kazienko, Katarzyna Musiał, and Aleksander Zgrzywa, 'Evaluation of Node Position Based on Email Communication," *Control and Cybernetics* 38, no. 1 (2009): 67–86.

42. It is not possible to provide definitive numbers here. In answer to an inquiry I made to FERC I was advised they were unable to supply precise figures for how many emails were originally removed from the Enron dataset in 2003. They further advised that the Enron dataset is currently under review and is not available to the public through their site.

43. Cohen, "Enron Email Dataset."

44. "Enron Email Data," Amazon Web Services, accessed June 11, 2017, https://aws.amazon.com/datasets/enron-email-data/.

45. "New EDRM Enron Email Data Set," *EDRM Duke Law*, accessed June 8, 2017, http://www.edrm.net/resources/data-sets/edrm-enron-email-data-set.

46. Ady Cassidy and Matthew Westwood-Hill, "Removing PII from the EDRM Enron Data Set," May 13, 2015, www.nuix.com/enron.

47. Jim McGann, "Index Engines Finds More Dirt on Nuix's 'Cleansed' Enron Data Set," *Index Engines* (blog), May 23, 2013, accessed June 11, 2017, http://www.poweroverinformation.com/index-engines-finds-more-dirt-on-nuixs-cleansed-enron-data-set/.

48. Michael S. Schmidt, "Hillary Clinton Used Personal Email Account at State Dept., Possibly Breaking Rules," *New York Times*, March 2, 2015, http://www.nytimes.com/2015/03/03/us/politics/hillary-clintons-use-of-private-email-at-state-department-raises-flags.html.

49. Rebecca Coates Nee and Mariana De Maio, "A 'Presidential Look'? An Analysis of Gender Framing in 2016 Persuasive Memes of Hillary Clinton," *Journal of Broadcasting & Electronic Media* 63, no. 2 (2019): 304–321.

50. US Department of Justice, "Report on the Investigation into Russian Interference in the 2016 Presidential Election," Volume I of II, Special Counsel Robert S. Mueller Washington, DC, March 2019, 1.

51. See, for example, Matthew J. Kushin, Masahiro Yamamoto, and Francis Dalisay, 'Societal Majority, Facebook, and the Spiral of Silence in the 2016 US Presidential Election," *Social Media + Society* (April–June 2019): 1–11; Gina Masullo Chen, Martin J. Riedl, Jeremy L. Shermak, Jordon Brown, and Ori Tenenboim, "Breakdown of Democratic Norms? Understanding the 2016 US Presidential Election through Online Comments," *Social Media + Society* (April–June 2019): 1–13; Regina G. Lawrence and Amber E. Boydstun, "What We Should Really Be Asking about Media Attention to Trump," *Political Communication* 34, no. 1 (2017): 150–153; and Bryan McLaughlin and Timothy Macafee, "Becoming a Presidential Candidate: Social Media Following and Politician Identification," *Mass Communication and Society* 22, no. 5 (2019): 584–603.

52. Amy Chozick and Steve Eder, "Membership in Clinton's Email Domain Is Remembered as a Mark of Status," *New York Times*, March 4, 2015, http://www.nytimes.com/2015/03/05/us/politics/membership-in-clintons-email-domain-is-remembered-as-a-mark-of-status.html.

53. "36 CFR 1236.22—What Are the Additional Requirements for Managing Electronic Mail Records?," Cornell Law School, accessed June 11, 2015, https://www.law.cornell.edu/cfr/text/36/1236.22.

54. "Updated: The Facts About Hillary Clinton's Emails," *The Briefing*, December 11, 2015, https://www.netadvisor.org/wp-content/uploads/2016/08/2015-08-23-Updated_-The-Facts-About-Hillary-Clintons-Emails-HRC.pdf.

55. "National Archives Releases State Department Letter Re: Email Recordkeeping," *National Archives*, 2015 press releases, accessed June 11, 2017, https://www.archives.gov/press/press-releases/2015/nr15-65.html.

56. Jason Baron, quoted in Eugene Kiely, "More Spin on Clinton Emails," FactCheck.org, September 8, 2015, http://www.factcheck.org/2015/09/more-spin-on-clinton-emails/.

57. Ari Fleischer, Twitter post, March 10, 2015, 12:34 p.m., https://twitter.com/AriFleischer/status/575379211407335424.

58. Reince Priebus, "Reince Priebus: 53 Questions I Have For Hillary Clinton," *Independent Journal Review*, accessed June 11, 2017, http://journal.ijreview.com/2015/03/242744-reince-priebus-53-questions-hillary-clinton/.

59. "State Department Releases over 3,000 Clinton Emails on New Year's Eve," *Fox News*, January 1, 2016, http://www.foxnews.com/politics/2016/01/01/state-department-releases-over-3000-clinton-emails-on-new-years-eve.html.

60. John Hayward, "Laws are for the Little People: State Dept. Violates Court Order on Clinton Emails," *Breitbart*, December 31, 2015, http://www.breitbart.com/big-government/2015/12/31/laws-little-people-state-dept-violates-court-order-clinton-emails/.

61. Cory Bennett, "Silence is Hurting Clinton, Dem Says," *The Hill* (blog), March 8, 2015, http://thehill.com/blogs/blog-briefing-room/234995-top-dem-silence-on-email-is-hurting-clinton.

62. M. J. Lee, "Hillary Clinton Donors Frustrated by Email Controversy," *CNN Politics*, March 11, 2015, http://edition.cnn.com/2015/03/10/politics/donors-hillary-clinton-emails.

63. Jamelle Bouie, "Why Democrats Have the Right to Be Angry with Clinton," *Slate*, September 1, 2015, http://www.slate.com/articles/news_and_politics/politics/2015/09/democrats_have_the_right_to_be_angry_with_hillary_clinton_her_email_scandal.html.

64. "Letter from National Archives, Chief Records Officer, Paul M. Wester to Deputy Assistant Secretary for Global Information Services, Margaret P. Grafield re: Possible Alienation of Former Secretary of State, Hillary Clinton's Federal Email Records, March 3, 2015," *National Archives*, 2015 press releases, http://www.archives.gov/press/press-releases/2015/pdf/nara-letter-to-state-department-3-3-15.pdf.

65. "State Department Response to National Archives' Letter re: State Department Emails," National Archives, 2015 press releases, March 3, 2015, http://www.archives.gov/press/press-releases/2015/nr15-65-state-dept-response-to-nara-re-email-recordkeeping.html.

66. "Letter to Hillary Clinton's representative, Cheryl Mills re: the Federal records Act of 1950, November 12, 2014," National Archives, 2015 press releases, http://www.archives.gov/press/press-releases/2015/pdf/attachment4-clinton-letter.pdf.

67. "Response to Under Secretary of State for Management, Patrick Kennedy from Hillary Clinton's Representative, Cheryl Mills, December 5, 2014," National Archives, 2015 press releases, http://www.archives.gov/press/press-releases/2015/nr15-65-state-dept-response-to-nara-re-email-recordkeeping.html.

68. "State Department's Response to National Archives Chief Records Officer, Paul M. Wester's March 3, 2015, Letter About Email Practices of Former Secretaries of State, April 2, 2015," National Archives, 2015 press releases, http://www.archives.gov/press/press-releases/2015/pdf/state-dept-response-to-wester-04-02-2015.pdf.

69. Eugene Kiely, "Clinton's Email Narrative, Interrupted," FactCheck.org, September 23, 2015, http://www.factcheck.org/2015/09/clintons-email-narrative-interrupted/.

70. Stephen Dinan, "Hillary Clinton Deleted 32,000 Private Emails, Refuses to Turn Over Server," *Washington Times*, March 10, 2015, http://www.washingtontimes.com/news/2015/mar/10/hillary-clinton-deleted-32000-private-emails-refus/.

71. "Updated: The Facts."

72. Zeke J. Miller, "Transcript: Everything Hillary Clinton Said on the Email Controversy," *Time*, March 11, 2015, http://time.com/3739541/transcript-hillary-clinton-email-press-conference/.

73. Ben Mathis-Lilley, "Hillary Clinton's 'Thorough' Email Review Was a Keyword Search," *Slate*, March 13, 2015, www.slate.com/blogs/the_slatest/2015/03/13/hillary_clinton_email_review_searched_for_some_words_deleted_everything.html.

74. Ed Morrissey, "New Story From Hillary Camp: Sure, We Read all 32,000 Deleted E-mails First," *Hot Air*, March 16, 2015, http://hotair.com/archives/2015/03/16/new-story-from-hillary-camp-sure-we-read-at-all-30000-deleted-e-mails-first/.

75. Shushannah Walshe and Liz Kreutz, "Hillary Clinton's Deleted Emails Were Individually Reviewed After All, Spokesman Says," *ABC News*, March 15, 2015, http://abcnews.go.com/Politics/hillary-clintons-deleted-emails-individually-reviewed-spokesman/story?id=29654638.

76. Brian Bennett and Evan Halper, "What You Need to Know About Hillary Clinton's Emails," *Los Angeles Times*, October 13, 2015, http://www.latimes.com/politics/la-na-pol-prez-clinton-emails-q-and-a-html-htmlstory.html.

77. Rachel Bade, "FBI Steps Up Probes in Clinton Email Probe," *Politico*, November 10, 2015, http://www.politico.com/story/2015/11/hillary-clinton-email-fbi-probe-215630.

78. Stephen Dinan, "State Department Can't Find Enough Staffers to Process Hillary Clinton's Emails," *Washington Times*, October 13, 2015, http://www.washingtontimes.com/news/2015/oct/13/hillary-clinton-emails-state-department-cant-find-/.

79. Laura Vorberg and Anna Zeitler, "'This Is (Not) Entertainment!': Media Constructions of Political Scandal Discourses in the 2016 US Presidential Election," *Media, Culture & Society* 41, no. 4 (2019): 417–432.

80. "Declaration of John F. Hackett," *Jason Leopold v. US Department of State*, Case 1:15-cv-00123-RC, Filed May 18, 2015 (US District Court for the District of Columbia, 2015), http://benghazi.house.gov/sites/republicans.benghazi.house.gov/files/State.Hackett.Declaration.2015.05.18.pdf.

81. David Fazekas, "Hillary Clinton's 55,000 Pages of Emails: A Whole Lot of Paper," *ABC News*, July 5, 2016, http://abcnews.go.com/Politics/hillary-clintons-55000-pages-emails-lot-paper/story?id=32663275.

82. "Updated: The Facts."

83. "Declaration of John F. Hackett."

84. Rosalind, S. Helderman, "State Department Releases Final Batch of Clinton Emails," *Washington Post*, February 29, 2016, https://www.washingtonpost.com/news/post-politics/wp/2016/02/29/state-department-releases-final-batch-of-clinton-emails/.

85. Kristina Wong, "Happy New Year! State Drops 5,500 Pages of Clinton Emails," *The Hill* (blog), December 31, 2015, http://thehill.com/policy/national-security/264527-feds-release-new-batch-of-clinton-emails-on-new-years-eve.

86. Laura Koran "State Department Releases New Batch of Clinton Emails", *CNN*, October 7, 2016, https://edition.cnn.com/2016/10/07/politics/clinton-emails-state-department/index.html.

87. Rachael Bade, "Judge Explodes Over Hillary Email Delays," *Politico*, July 30, 2015, http://www.politico.com/story/2015/07/judge-explodes-over-hillary-email-delays-120804.

88. Bade, "Judge Explodes Over Hillary Email Delays."

89. Hillary Clinton, Twitter post, March 4, 2015, 8:35 p.m., https://twitter.com/HillaryClinton/status/573340998287413248.

90. Dan Roberts, "Hillary Clinton Breaks Media Silence and Insists: 'I Want Those Emails Out,'" *Guardian*, May 19, 2015, http://www.theguardian.com/us-news/2015/may/19/hillary-clinton-emails-media-iowa.

91. Dan Roberts, "Hillary Clinton Foists Private Email Release Delays on Obama Administration," *Guardian*, July 30, 2015, http://www.theguardian.com/us-news/2015/jul/30/hillary-clinton-emails-release-delay-state-department.

92. Lauren Carroll, "Hillary Clinton: Emails Wouldn't be Public if I Hadn't Asked For It," *Politifact*, August 20, 2015, http://www.politifact.com/truth-o-meter/statements/2015/aug/20/hillary-clinton/hillary-clinton-emails-wouldnt-be-public-if-i-hadn/.

93. Dan Merica, "Clinton Used iPad for Personal Email at State," *CNN Politics*, March 31, 2015, http://edition.cnn.com/2015/03/31/politics/hillary-clinton-ipad-e-mail-devices/.

94. Gregory Korte and Edward C. Baig, "Could Clinton's BlackBerry Handle Multiple E-mail Accounts?" *USA Today*, March 10, 2015, http://www.usatoday.com/story/news/politics/2015/03/10/hillary-clinton-emails-blackberry/24725993/.

95. Jon Favreau, Twitter post, March 10, 2015, 12:05 p.m., https://twitter.com/jonfavs/status/575371820892729344.

96. John Podhoretz, Twitter post, March 10, 2015, 12:06 p.m., https://twitter.com/jpodhoretz/status/575372052573589504.

97. Jon Favreau, Twitter post, March 10, 2015, 12:06 p.m., https://twitter.com/jonfavs/status/575372209964781568.

98. Evan McMorris-Santoro, "Former Administration Aides: Expectation Was Disclosure For Personal Email," *BuzzFeed*, March 6, 2015, http://www.buzzfeed.com/evanmcsan/obama-administration-private-email#.vqG5nq2yn.

99. Josh Voorhees, "A Crystal Clear Explanation of Hillary's Confusing Email Scandal," *The Slatest*, August 20, 2015, http://www.slate.com/blogs/the_slatest/2015/08/20/hillary_clinton_email_scandal_explained.html.

100. Symantec, "2018 Shadow Data Report," https://www.uk.insight.com/content/dam/insight-web/en_GB/Buy/shops/symantec/Symc_IR_ShadowDataReport-2018_en_v1b.pdf.

101. Tom Kaneshige, "Hillary Clinton is Now the Face of Shadow IT," *CIO*, March 12, 2015, http://www.cio.com/article/2895805/it-management/hillary-clinton-is-now-the-face-of-shadow-it.html.

102. Ramon Lobato, *Shadow Economies of Cinema: Mapping Informal Film Distribution* (London: British Film Institute/Palgrave Macmillan, 2012); Finn Brunton, *Spam: A Shadow History of the Internet* (Cambridge, MA: MIT Press, 2013).

103. Steffi Haag, Andreas Eckhardt, and Andrew Schwarz, "The Acceptance of Justifications among Shadow IT Users and Nonusers—An Empirical Analysis," *Information & Management* 56, no. 5 (July 2019): 731–741.

104. "848 Pages Tagged," *Wall Street Journal*, May 22, 2015, http://graphics.wsj.com/hillary-clinton-emails/.

105. "Hillary Clinton's Benghazi Emails—Live Dive," *Wall Street Journal*, May 22, 2015, http://blogs.wsj.com/washwire/2015/05/22/hillary-clintons-benghazi-emails-live-dive/.

106. "Hillary Clinton's Emails: Uncover the Political Landscape in Hillary Clinton's Emails," *kaggle*, accessed June 11, 2017, https://www.kaggle.com/kaggle/hillary-clinton-emails.

107. Hanna Julienne, "World Geopolitics through Hillary Clinton's Emails," *Dataiku*, October 25, 2015, http://www.dataiku.com/blog/2015/10/26/t-hillary.html; Christopher Salahub and Wayne Oldford, "About 'Her Emails,'" The Royal Statistical Society, *Significance* 15, no. 3 (June 2018): 34–37, https://doi.org/10.1111/j.1740-9713.2018.01148.x.

108. Kenneth Goldsmith and Francesco Urbano Ragazzi (curator), *HILLARY: The Hillary Clinton Emails* (exhibition), Zuecca Projects 2019, http://www.zueccaprojects.org/project/hillary/.

109. Sarah Cascone, "Hillary Clinton's Emails Are on View in an Art Show in Venice," *ARTnet News*, September 12, 2019, https://news.artnet.com/art-world/hillary-clinton-reads-emails-venice-art-show-1648867.

Chapter 7: The Art of Email

1. Robert Mitchell, *Bioart and the Vitality of Media* (Seattle: University of Washington Press, 2010).

2. Mitchell, *Bioart and the Vitality of Media*, 5–6.

3. Mitchell, 8.

4. Darren Tofts and Murray McKeich, *Memory Trade: A Prehistory of Cyberculture.* North Ryde, NSW: 21C/Interface, 1998; Oliver Grau, *Virtual Art: From Illusion to Immersion* (Cambridge, MA: MIT Press, 2003); Michelle Henning, *Museums, Media and Cultural Theory* (Maidenhead, UK, and New York: Open University Press, 2006); Anna Munster, *Materializing New Media: Embodiment in Information Aesthetics* (Hanover, NH: Dartmouth College Press, 2006); Beryl Graham (ed.), *New Collecting: Exhibiting and Audiences after New Media Art* (New York: Routledge, 2014).

5. Victoria Vesna (ed.), *Database Aesthetics: Art in the Age of Information Overflow* (Minneapolis: University of Minnesota Press, 2007).

6. Gabriella Giannachi, *Archive Everything: Mapping the Everyday* (Cambridge, MA: MIT Press, 2017).

7. Susan Leigh Star, "This is Not a Boundary Object: Reflections on the Origin of a Concept," *Science, Technology, & Human Values* 35, no. 5 (2010): 601–617.

8. Nan McCarthy, *Chat: A Cybernovel* (Berkeley: Peachpit Press, 1996); Avodah Offit, *Virtual Love: A Novel*, (New York: Simon & Schuster, 1994); Stephanie Fletcher, *E-Mail: A Love Story* (London: Headline Book Publishing, 1996); Linda Burgess and Stephen Stratford, *Safe Sex: An E-mail Romance* (Auckland: Godwit Publishing Limited, 1997); Jessica Adams *Single White E-mail* (London: Black Swan, 1999); Sylvia Brownrigg, *Metaphysical Touch* (New York: Farrar, Straus and Giroux, 1999). See also: Esther Milne, "WriteSites: The Email Liaison as Novel," *RealTime* 38 (August–September 2000). For an overview of later email novels see: Jill Walker Rettberg, "Email Novels," in *The Johns Hopkins Guide to Digital Media*, ed. Marie-Laure Ryan, Lori Emerson, and Benjamin Robertson (Baltimore: Johns Hopkins University Press, 2014).

9. Kathy Acker and McKenzie Wark, *I'm Very Into You: Correspondence 1995–1996*, ed. Matias Viegener, afterword by John Kinsella (New York: Semiotext(e), 2015).

10. Acker and Wark, *I'm Very Into You*, 5.

11. The rise of the epistolary genre like the rise of the novel itself is the site of energetic critical debate. For key works see: Janet Gurkin Altman, *Epistolarity: Approaches to a Form* (Columbus: Ohio State Universtiy Press, 1982); Thomas O Beebee, *Epistolary Fiction in Europe: 1500–1850* (Cambridge: Cambridge University Press, 2006); Kym Brindle, *Epistolary Encounters in Neo-Victorian Fiction: Diaries and Letters* (New York: Palgrave Macmillan, 2013); Mary A. Favret, *Romantic Correspondence: Women, Politics and the Fiction of Letters* (Cambridge: Cambridge University Press, 2007); Catherine Golden, *Posting It: The Victorian Revolution in Letter Writing* (Gainesville: University Press of Florida, 2010); Ruth Perry, *Women, Letters, and the Novel* (New York: AMS Press, 1980); Nicola J. Watson, *Revolution and the Form of the British Novel, 1790–1825: Intercepted Letters, Interrupted Seductions* (Oxford: Clarendon Press, 2001).

12. Laura Rotunno, *Postal Plots in British Fiction, 1840–1898: Readdressing Correspondence in Victorian Culture* (New York: Palgrave Macmillan, 2013), 1–2.

13. Richard Menke, *Telegraphic Realism: Victorian Fiction and Other Information Systems* (Stanford, CA: Stanford University Press, 2008), 6.

14. Menke, *Telegraphic Realism*, 3–4.

15. Thomas Pynchon, *The Crying of Lot 49* (Philadelphia: Lippincott, 1966).

16. Linda Kauffman, *Special Delivery: Epistolary Modes in Modern Fiction* (Chicago: University of Chicago Press, 1992).

17. Chuck Welch, ed., *Eternal Network: A Mail Art Anthology* (Calgary, AB: University of Calgary Press: 1994), xviii.

18. Doris L. Bell, *Contemporary Art Trends 1960–1980: A Guide to Sources* (Metuchen, NJ: Scarecrow Press, 1981).

19. Jonathan Watkins and René Denizot, *On Kawara* (London: Phaidon, 2002).

20. John Held, "The Mail Art Exhibition: Personal Worlds to Cultural Strategies," in *At a Distance: Precursors to Art and Activism on the Internet*, ed. Annmarie Chandler and Norie Neumark (Cambridge, MA: MIT Press, 2005), 96.

21. Quoted in John Held, "The Mail Art Exhibition: Personal Worlds to Cultural Strategies," in *At a Distance: Precursors to Art and Activism on the Internet*, ed. Annmarie Chandler and Norie Neumark (Cambridge, MA: MIT Press, 2005), 96–97.

22. John Held, *Small Scale Subversion: Mail Art & Artistamps* (Breda, Netherlands: TAM Publishing, 2015), 164.

23. Jacobs quoted in Held, *Small Scale Subversion: Mail Art & Artistamps*, 165.

24. John Held, "Living Thing in Flight: Contributions and Liabilities of Collecting and Preparing Contemporary Avant-Garde Materials for an Archive," *Archives of American Art Journal* 40, no. 3/4 (2000): 10.

25. "Exhibition Listings," *Preview*, accessed June 11, 2017, http://www.preview-art.com/previews/02-2010/mailart.html.

26. "Lomholt Mail Art Archive," accessed June 11, 2017, http://www.lomholtmailartarchive.dk/intro.

27. "About Postcrossing," Postcrossing, accessed June 11, 2017, https://www.postcrossing.com/about.

28. "What is Mailart?" Mailart 365, accessed June 11, 2017, http://www.mailart365.com/what-is-mailart-365/.

29. "andysalterego," Mailart 365, accessed June 11, 2017, http://www.mailart365.com/author/andysalterego/.

30. "World's Smallest Post Service," accessed September 28, 2019, https://www.leafcutterdesigns.com/worlds-smallest-post-service/; "Our Shadow Selves from Canada," *Dear You*, accessed September 28, 2019, http://www.dearyouartproject

.com/; "External Heart Drive," accessed September 28, 2019, https://external-heart-drive.tumblr.com/; Postcard from the Past: "The Funniest, Weirdest and Most Moving Real Messages from the Backs of Old Postcards," 2019, http://postcardfromthepast.co.uk/; The Edwardian Postcard Project, https://www.facebook.com/edwardianpostcardsproject; Letters of Note, http://www.lettersofnote.com/.

31. *PostSecret*, accessed June 11, 2017, http://postsecret.com/.

32. For an extended discussion see Jenny Kennedy and Esther Milne, "Public Privacy: Reciprocity and Silence," *Platform: Journal of Media and Communication* 4, no. 2 (2013).

33. Mailart twitter home, https://twitter.com/mailart365; Postcrossing twitter home, https://twitter.com/postcrossing.

34. pooi_chin Instagram, https://www.instagram.com/pooi_chin/.

35. Tom McCormack, "Emoticon, Emoji, Text II: Just ASCII," *Rhizome* (blog), April 30, 2013, http://rhizome.org/editorial/2013/apr/30/emoticon-emoji-text-ii-ascii/.

36. Alan Riddell, ed., *Typewriter Art* (London: London Magazine Editions, 1975), 15.

37. Dirk Krecker quoted in Barrie Tullett, *Typewriter Art: A Modern Anthology* (London: Laurence King Publishing, 2014), 127.

38. Marvin Sackner and Ruth Sackner, *The Art of Typewriting* (London: Thames & Hudson, 2015), xx.

39. "The Sackner Archive of Concrete and Visual Poetry," accessed June 11, 2017, http://ww3.rediscov.com/sacknerarchives/Welcome.aspx?518201785314.

40. Brenda Danet, *Cyberpl@y: Communicating Online* (New York: Bloomsbury: 2001), 39.

41. Danet, *Cyberpl@y*, 210–212.

42. "You've Got Mail," accessed June 11, 2017, http://youvegotmail.warnerbros.com/cmp/4frameset.html.

43. Danet, *Cyberpl@y*, 209–210.

44. Menno Pieters, "Signature Museum: ASCII Email Signature Collection," September 2012, http://signaturemuseum.pieters.cx/;

45. For example see: Ryan M. Milner, *The World Made Meme Public Conversations and Participatory Media* (Cambridge, MA: MIT Press, 2016); Kate Miltner, "From #Feels to Structure of Feeling: The Challenges of Defining 'Meme Culture,'" in Festival of Memeology, *Culture Digitally*, October 29, 2015, http://culturedigitally.org/2015/10/memeology-festival-02-from-feels-to-structure-of-feeling-the-challenges-of-defining-meme-culture/; Linda K. Börzsei, "Makes a Meme Instead: A Concise History of Internet Memes," *New Media Studies Magazine* 7, (2013): 152–189.

46. Linda Kauffman, *Discourses of Desire: Gender, Genre, and Epistolary Fictions* (Ithaca: Cornell University Press, 1986); Elizabeth Heckendorn Cook, *Epistolary Bodies: Gender*

and Genre in the Eighteenth-Century Republic of Letters (Stanford, CA: Stanford University Press 1996); Elizabeth C. Goldsmith, ed., *Writing the Female Voice* (Boston: Northeastern University Press, 1989); Amanda Gilroy and W. M. Verhoeven, *Epistolary Histories: Letters, Fiction, Culture* (Charlottesville: University Press of Virginia, 2000); Katie-Louise Thomas, *Postal Pleasures: Sex, Scandal, and Victorian Letters* (New York: Oxford University Press, 2012); Anne Bower, *Epistolary Responses: The Letter in 20th-century American Fiction and Criticism* (Tuscaloosa: University of Alabama Press, 2014).

47. Carl Steadman, "Two Solitudes [an e-mail romance]," 1994. Steadman's own copy.

48. Joe Fasbinder, "Virtual Love and Cyber-Romance," *UPI*, February 12, 1995, http://www.upi.com/Archives/1995/02/12/Virtual-love-and-cyber-romance/9702792565200/.

49. Carl Steadman, "Two Solitudes," *InterText* 5, no. 1 (1995), http://intertext.com/magazine/v5n1/solitudes.html.

50. Steadman, "Two Solitudes."

51. Steadman, "Two Solitudes."

52. "We Think Alone: A Project by Miranda July," press release, *Magasin III*, https://www.magasin3.com/en/pressrelease/we-think-alone-a-project-by-miranda-july-2/.

53. "What Is Immersion?" *Immersion*, accessed June 11, 2017, https://immersion.media.mit.edu/.

54. "What Is Immersion?"

55. Lisa Main, "Data Retention: What is Metadata and How Will It Be Defined by New Australian Laws?" *ABC News*, March 17, 2015, http://www.abc.net.au/news/2015-03-17/what-is-metadata-how-will-it-be-defined-by-new-australia-laws/6325908.

56. Quoted in David Cole, "We Kill People Based on Metadata," *New York Review of Books*, May 10, 2014, http://www.nybooks.com/daily/2014/05/10/we-kill-people-based-metadata/.

57. José van Dijck, "Datafication, Dataism and Dataveillance: Big Data Between Scientific Paradigm and Ideology," *Surveillance & Society* 12, no. 2 (2014): 197–208.

58. César Hidalgo interviewed in Abraham Riesman, "What Your Metadata Says About You," *Boston Globe*, June 30, 2013, https://www.bostonglobe.com/ideas/2013/06/29/what-your-metadata-says-about-you/SZbsH6c8tiKtdCxTdl5TWM/story.html.

59. Cesar A. Hidalgo, "What I Learned from Visualizing Hillary Clinton's Emails," MIT Media Lab, Nov 5, 2016, https://medium.com/mit-media-lab/what-i-learned-from-visualizing-hillary-clintons-leaked-emails-d13a0908e05e#.72rtwz8tt.

60. Brian Fuata, "All Titles, No Centre Crux: The Email Performances in PERFORMA 2015," *PERFORMA 15,* http://15.performa-arts.org/events/all-titles-no-centre-crux-the-email-performances-in-performa-2015.

61. "Small Scale Performance Propositions," *Realtime*, accessed June 11, 2017, http://www.realtimearts.net/studio-artist/small-scale-performance-propositions.

62. Jo Thomas, "For Inventor of Eudora, Great Fame, No Fortune," *New York Times*, January 21, 1997 http://www.nytimes.com/library/cyber/week/012197eudora.html.

63. "Informed Delivery," https://informeddelivery.usps.com/box/pages/intro/start.action (thanks to Rowan Wilken for alerting me to this service).

Conclusion

1. Rafael Epstein, "Changing Tracks," *Drive* (radio broadcast) ABC, Radio Melbourne, 2019, https://www.abc.net.au/radio/melbourne/programs/drive/changing-tracks/.

2. PRX The Public Radio Exchange, "Sample Pitches," *This American Life* (radio broadcast), https://www.thisamericanlife.org/about/sample-pitches.

3. Ben Highmore, *Everyday Life and Cultural Theory* (New York: Taylor and Francis, 2002), 115.

4. Andrew L. Russell, *Open Standards and the Digital Age: History, Ideology, and Networks* (New York: Cambridge University Press, 2014).

5. For example see: Mar Hicks, *Programmed Inequality How Britain Discarded Women Technologists and Lost Its Edge in Computing* (Cambridge, MA: MIT Press: 2017).

6. Susan J Douglas, "Some Thoughts on the Question "How Do New Things Happen?'" *Technology and Culture* 51, no. 2 (April 2010): 295.

7. Jean Burgess and Axel Bruns, "Easy Data, Hard Data: The Politics and Pragmatics of Twitter Research after the Computational Turn," in *Compromised Data: From Social Media to Big Data* , ed. G. Langlois, J. Redden, and G. Elmer (London: Bloomsbury Publishing, 2015), 93–111; Nicholas A. John and Asaf Nissenbaum, "An Agnotological Analysis of Apis: Or, Disconnectivity and the Ideological Limits of Our Knowledge of Social Media," *The Information Society* 35, no. 1 (2019): 1–12, https://doi.org/10.1080/01972243.2018.1542647.

8. Kenzie Burchell, "Everyday Communication Management and Perceptions of Use," *Convergence* 23, no. 4 (2017): 409–424.

9. Annalee Newitz, "Email Is the New Generation Gap," *Gizmodo*, November 29, 2015, https://www.gizmodo.com.au/2015/11/email-is-the-new-generation-gap/.

10. Danial Jameel, "In a World Where Email No Longer Exists," *TechCrunch*, March 6, 2015, https://techcrunch.com/2015/03/06/in-a-world-where-email-no-longer-exists/.

11. Newitz, "Email Is the New Generation Gap."

12. Newitz, "Email Is the New Generation Gap."

13. Susan Chun and Dale Kronkright, "Museums and the Web Deep Dive: Assessing Tools and Best Practices for Email Preservation and Access in Art Museums," Museums and the Web Conference, April 2–5, 2014, Baltimore, Maryland, http://mw2014.museumsandtheweb.com/museums-and-the-web-deep-dive-assessing-tools-and-best-practices-for-email-preservation-and-access-in-art-museums/.

14. Hyunmo Kang, Catherine Plaisant, Tamer Elsayed, and Douglas Oard, "Making Sense of Archived E-mail: Exploring the Enron Collection With NetLens," *Journal of the American Society for Information Science and Technology* 61, no. 4 (2009): 723–744.

15. "About Letters of Note," *Letters of Note*, accessed June 11, 2017, http://www.lettersofnote.com/.

16. "The Browning Letters," Baylor University Archives, accessed June 11, 2017, http://digitalcollections.baylor.edu/cdm/landingpage/collection/ab-letters.

17. William P. Vinh-Doyle, "Appraising Email (Using Digital Forensics): Techniques and Challenges," *Archives and Manuscripts* 45, no. 1 (2017): 18.

18. Vinh-Doyle, "Appraising Email," 20.

19. Andrew Waugh, "Email—A Bellwether Records System," *Archives and Manuscripts* 42, no. 2 (2014): 215.

20. Matthew G Kirschenbaum, Richard Ovenden, and Gabriela Redwine, *Digital Forensics and Born-Digital Content in Cultural Heritage Collections* (Washington, DC: Council on Library and Information Resources, 2010), https://www.clir.org/pubs/reports/reports/pub149/pub149.pdf.

21. "BitCurator NLP Project," *BitCurator*, UNC School of Information and Library Science, accessed June 12, 2017, https://www.bitcurator.net/bitcurator-nlp/.

22. Jeremy Leighton John, *Digital Forensics and Preservation* (Digital Preservation Coalition in association with Charles Beagrie Ltd., 2012), http://www.dpconline.org/docman/technology-watch-reports/810-dpctw12-03-pdf/file.

23. Chun and Kronkright, "Museums and the Web Deep Dive."

24. "The Collaborative Electronic Records Project," The Rockefeller Archive Center, accessed June 12, 2017, http://siarchives.si.edu/cerp/index.htm.

25. "Task Force on Technical Approaches to Email Archives," accessed June 11, 2017, http://www.emailarchivestaskforce.org/about/task-force-charge/.

26. Janet Abbate, *Inventing the Internet* (Cambridge, MA: MIT Press, 1999): 111.

27. Donna Haraway, *Modest_Witness@Second_Millennium. FemaleMan©_Meets_OncoMouseTM: Feminism and Technoscience* (New York: Routledge, 1997).

28. Richard Ling, *Taken for Grantedness: The Embedding of Mobile Communication into Society* (Cambridge, MA: MIT Press, 2012).

Bibliography

"36 CFR 1236.22—What Are the Additional Requirements for Managing Electronic Mail Records?" Cornell Law School. Accessed June 11, 2015, https://www.law.cornell.edu/cfr/text/36/1236.22.

"About Letters of Note." Letters of Note. Accessed June 11, 2017. http://www.lettersofnote.com/.

"About Postcrossing." Postcrossing. Accessed June 11, 2017. https://www.postcrossing.com/about.

Aamoth, Doug. "The Man Who Invented Email." *Time*, November 15, 2011.

Abbate, Janet. *Inventing the Internet*. Cambridge, MA: MIT Press, 1999.

Acker, Joan. "Hierarchies, Jobs, Bodies: A Theory of Gendered Organizations." *Gender and Society* 4, no. 2 (1990): 151–152.

Acker, Kathy, and McKenzie Wark. *I'm Very Into You: Correspondence 1995–1996*. Edited by Matias Viegener. New York: Semiotext(e), 2015.

Acland, Charles R., ed. *Residual Media*. Minneapolis: University of Minnesota Press, 2007.

Ad Council. "I am a Witness Campaign." Accessed April 9, 2017. http://iwitnessbullying.org/about.Adams, Jessica. *Single White E-mail*. London: Black Swan, 1999.

Adams, Richard. "Gawker's Christine O'Donnell Sex Smear Marks a New Low." *Guardian*, October 30, 2010. https://www.theguardian.com/world/richard-adams-blog/2010/oct/29/christine-odonnell-gawker-one-night-stand-dustin-dominiak.

Aguilar, Mario. "Internet Pioneers Slam $750,000 Settlement for the 'Man Who Invented Email.'" *Gizmodo*, November 4, 2016.

Ahmed, Sara. "Open Forum Imaginary Prohibitions: Some Preliminary Remarks on the Founding Gestures of the 'New Materialism.'" *European Journal of Women's Studies* 15, no. 1 (2008): 23–39.

Alt, Matt. *The Secret Lives of Emoji: How Emoticons Conquered the World*, Amazon Digital Services, Kindle Edition, 2016.

Altman, Janet Gurkin. *Epistolarity: Approaches to a Form*. Columbus: Ohio State University Press, 1982.

American Institute for Conservation of Historic and Artistic Works. "Giving to Cool." http://www.conservation-us.org/our-organizations/foundation-(faic)/initiatives/conservation-online-(cool)#.WOiZsfl97cc.

American Psychological Association. "Americans Stay Connected to Work on Weekends, Vacation and Even When Out Sick." September 4, 2013.

andysalterego. Mailart 365, accessed June 11, 2017. http://www.mailart365.com/author/andysalterego/.

Antonioua, Alexandros, and Dimitris Akrivosb. "Hulk Hogan and the Demise of Gawker Media: Wrestling with Problems of Celebrity Voyeurism, Newsworthiness and Tabloidization." *Journal of Media Law* 8, no. 2 (2016): 153–172.

Arana, Gabriel. "Gawker's Outing of Condé Nast's CFO Is Gay-Shaming, Not Journalism." *Huffington Post*, July 18, 2015. http://www.huffingtonpost.com.au/entry/gawker-conde-nast-david-geithner_us_55a90c56e4b0c5f0322d0b2c.

Ashcraft, Karen Lee. "Organized Dissonance: Feminist Bureaucracy as Hybrid Form." *Academy of Management* 44, no. 6 (2001): 1301–1322.

"Atos Boss Thierry Breton Defends his Internal Email Ban." *BBC News*, December 6, 2011. http://www.bbc.com/news/technology-16055310.

Attfield, Judy. *Wild Things: The Material Culture of Everyday Life*. Oxford: Berg, 2005.

Ayyadurai, V.A. Shiva. *The Email Revolution: The Power to Connect*. New York: Allworth Press, 2013.

Bade, Rachael. "Judge Explodes Over Hillary Email Delays." *Politico*, November 15, 2015. http://www.politico.com/story/2015/07/judge-explodes-over-hillary-email-delays-120804#ixzz3wQH6czry.

Bardini, Thierry. *Bootstrapping: Douglas Engelbart, Coevolution, and the Origins of Personal Computing*. Stanford, CA: Stanford University Press, 2000.

Baron, Naomi S. *Alphabet to Email: How Written English Evolved and Where It's Heading*. London: Routledge, 2000.

Baron, Naomi S. "Why Email Looks like Speech: Proofreading, Pedagogy and Public Face." In *New Media Language*, edited by Jean Aitchison and Diana M. Lewis. London: Routledge, 2003.

Bartling, Joe. "The Enron Data Set—Where Did It Come From." Bartling Forensic LLC. Accessed June 7, 2017. http://www.bartlingforensic.com/the-enron-data-set-where-did-it-come-from/.

Beebee, Thomas O. *Epistolary Fiction in Europe: 1500–1850*. Cambridge: Cambridge University Press, 2006.

Bekkerman, Ron, Andrew McCallum, and Gary Huang. 2004. "Automatic Categorization of Email into Folders: Benchmark Experiments on Enron and SRI Corpora." Center for Intelligent Information Retrieval, 2004.

Belkin, Liuba Y., William J. Becker, and Samantha A. Conroy. "Exhausted, But Unable to Disconnect: After-Hours Email, Work-Family Balance and Identification." *Academy of Management Proceedings*, 2016. https://doi.org/10.5465/AMBPP.2016.10353abstract.

Bell, Doris L. *Contemporary Art Trends 1960–1980: A Guide to Sources*. Metuchen, NJ: Scarecrow Press, 1981.

Bennett, Brian, and Evan Halper. "What You Need to Know About Hillary Clinton's Emails." *Los Angeles Times*, October 13, 2015. https://www.latimes.com/politics/la-na-pol-prez-clinton-emails-q-and-a-html-htmlstory.html.

Bennett, Cory. "Silence is Hurting Clinton, Dem Says." *The Hill* (blog), November 8, 2015. http://thehill.com/blogs/blog-briefing-room/234995-top-dem-silence-on-email-is-hurting-clinton.

Berge, Zane L., and Mauri P. Collins. "Perceptions of E-moderators about Their Roles and Functions in Moderating Electronic Mailing Lists." *Distance Education* 21, no. 1 (2000): 81–100.

Berners-Lee, Mike. *How Bad Are Bananas? The Carbon Footprint of Everything*. London: Profile Books, 2010.

Berry, Michael, and Murray Browne. "Email Surveillance Using Non-negative Matrix Factorization." *Computational & Mathematical Organization Theory* 11 (2005): 249–264.

Billing, Yvonne Due. "Gender Equity—A Bureaucratic Enterprise?" In *The Values of Bureaucracy*, edited by Paul du Gay. Oxford: Oxford University Press, 2005.

"BitCurator NLP Project." UNC School of Information and Library Science. Accessed June 12, 2017. https://www.bitcurator.net/bitcurator-nlp/.

Bolkan, San, and Jennifer Linn Holmgren. "'You Are Such a Great Teacher and I Hate to Bother You But …': Instructors' Perceptions of Students and their use of Email Messages with Varying Politeness Strategies." *Communication Education* 61 no. 3 (2012): 253–270.

Bollmer, Grant. 2015. "Technological Materiality and Assumptions About 'Active' Human Agency." *Digital Culture and Society* 1, no. 1 (2015): 95–110.

Bolton, Whitney. "CMC and Email: Casting a Wider Net." *English Today* 7, no. 4 (1991): 34–38.

Borenstein, Nathaniel. "Part # 146: We're Slaves to Our Attachments." *Mimecast* (blog), March 25, 2011. https://www.mimecast.com/blog/2011/03/part-146-were-slaves-to-our-attachments/.

Borger, Julian. "Tapes Reveal Enron's Secret Role in California's Power Blackouts." *Guardian*, February 5, 2005. http://www.theguardian.com/business/2005/feb/05/enron.usnews.

Börzsei, Linda K. "Makes a Meme Instead: A Concise History of Internet Memes." *New Media Studies Magazine* 7 (2013): 152–189.

Bosworth, Martin. "The Death of E-Mail?" *Consumer Affairs*, April 7, 2006. www.consumeraffairs.com/news04/2006/07/email_fading.html.

Botan, Carl. "Communication Work and Electronic Surveillance: A Model for Predicting Panoptic Effects." *Communication Monographs* 63, no. 4 (1996): 293–313.

Bott, Ed. "What Does the New Gmail API Mean for Internet Standards?" *ZDNet*, June 26, 2014.

Bouie, Jamelle. "Why Democrats Have the Right to Be Angry With Clinton." *Slate*, September 1, 2015, http://www.slate.com/articles/news_and_politics/politics/2015/09/democrats_have_the_right_to_be_angry_with_hillary_clinton_her_email_scandal.html.

Bourgault, Sophie. "Prolegomena to a Caring Bureaucracy," *European Journal of Women's Studies* 24, no. 3 (2016): 202–217. https://doi.org/10.1177/1350506816643730.

Bower, Anne. *Epistolary Responses: The Letter in 20th-century American Fiction and Criticism*. Tuscaloosa: University of Alabama Press, 2014.

Bradner, Scott. IETF Working Group Guidelines and Procedures, BCP 25, RFC 2418. https://doi.org/10.17487/RFC2418, September 1998.

Braman, Sandra. "The Interpenetration of Technical and Legal Decision-Making for the Internet." *Information, Communication & Society* 13, no. 3 (2010), 310.

Braman, Sandra. "Privacy by Design: Networked Computing, 1969–1979." *New Media & Society* 14, no. 5 (2012), 799.

Brewer, Lynn, and Matthew Scott Hansen, *Confessions of an Enron Executive: A Whistleblower's Story*. Bloomington, IL: AuthorHouse: 2004.

Brindle, Kym. *Epistolary Encounters in Neo-Victorian Fiction: Diaries and Letters*. New York: Palgrave Macmillan, 2013.

Brodie, Ian. *Email Persuasion: Captivate and Engage Your Audience, Build Authority and Generate More Sales with Email Marketing*: Myrtle Beach, SC: Rainmaker, 2013.

Brooks, Peter. *Reading for the Plot: Design and Intention in Narrative*. Cambridge, MA: Harvard University Press, 1984.

Brownrigg, Sylvia. *Metaphysical Touch*. New York: Farrar, Straus and Giroux, 1999.

Brügger, Niels. "Australian Internet Histories: Past, Present and Future: An Afterword." *Media International Australia, theme issue: Internet Histories* 143 (2012): 159–165.

Brunton, Finn. *Spam: A Shadow History of the Internet.* Cambridge, MA: MIT Press, 2013.

Bugeaud, Theresa, and Jonathan Benton. "Benchmarks: The 2018 Report of How Nonprofits are Performing Online," M&R and Nonprofit Technology Network, 2018.

Bulkeley, William M. "Echomail Provides an Answer for the Avalanche of E-mail." *Wall Street Journal,* November 15, 2001. http://www.wsj.com/articles/SB100577687 3756157960.

Bunker, Matthew D., Paul H. Gates Jr., and William C. Nevin. Anti-SLAPP Statutes Offer Tool for Media Defendants. *Newspaper Research Journal* 35, no. 1 (2014): 6–19.

Burchell, Kenzie. "Everyday Communication Management and Perceptions of Use." *Convergence* 23, no. 4 (2017): 409–424.

Burgess, Jean, and Axel Bruns. "Easy Data, Hard Data: The Politics and Pragmatics of Twitter Research after the Computational Turn." In *Compromised Data: From Social Media to Big Data,* edited by G. Langlois, J. Redden, and G. Elmer, 93–111. London: Bloomsbury Publishing, 2015.

Burgess, Linda, and Stephen Stratford. *Safe Sex: An E-mail Romance.* Aukland: Godwit Publishing Limited, 1997.

Carey, John, and Martin C. J. Elton. "The Other Path to the Web: The Forgotten Role of Videotex and Other Early Online Services." *New Media Society* (2009): 241–260.

Carley, Kathleen, and David Skillicorn, eds. "Special Issue on Analyzing Large Scale Networks: The Enron Corpus." *Computational & Mathematical Organization Theory* 11 (2005): 179–181.

Carpenter, Brian E., and Craig Partridge. "Internet Requests for Comments (RFCs) as Scholarly Publications." *ACM SIGCOMM Computer Communication Review* 40, no. 1 (2010).

Carroll, Lauren. "Hillary Clinton: Emails Wouldn't Be Public If I Hadn't Asked for It." *Politifact,* August 20, 2015. http://www.politifact.com/truth-o-meter/statements /2015/aug/20/hillary-clinton/hillary-clinton-emails-wouldnt-be-public-if-i-hadn/.

Cásarez, Nicole. "Electronic Mail and Employee Relations: Why Privacy Must Be Considered." *Public Relations Quarterly* 37, no. 2 (1992): 37–40.

Cassidy, Ady, and Matthew Westwood-Hill 2015. "Removing PII from the EDRM Enron Data Set." May 13, 2015. www.nuix.com/enron.

Castellucci, Steven J., and Scott MacKenzie. "Gathering Text Entry Metrics on Android Devices." International Conference on Multimedia and Human-Computer Interaction—MHCI 2013, Ottawa, Canada.

Chen, Gina Masullo, Martin J. Riedl, Jeremy L. Shermak, Jordon Brown, and Ori Tenenboim. "Breakdown of Democratic Norms? Understanding the 2016 US Presidential Election through Online Comments." *Social Media + Society* (April–June 2019): 1–13.

Chessum, Jake. "Want to Prove Patience Pays Off? Ask the Founders of This 17-Year-Old $525 Million Email Empire!" *Inc.*, Winter 2017. https://www.inc.com/magazine/201802/mailchimp-company-of-the-year-2017.html.

Chozick, Amy, and Steve Eder. "Membership in Clinton's Email Domain Is Remembered as a Mark of Status." *New York Times*, March 4, 2015. http://www.nytimes.com/2015/03/05/us/politics/membership-in-clintons-email-domain-is-remembered-as-a-mark-of-status.html.

Chun, Susan, and Dale Kronkright. "Museums and the Web Deep Dive: Assessing Tools and Best Practices for Email Preservation and Access." Museums and the Web Conference, April 2–5, 2014, Baltimore, MD, USA.

Chun, Wendy Hui Kyong. "Introduction." In *New Media, Old Media: A History and Theory Reader*, edited by Wendy Hui Kyong Chun and Thomas Keenan. New York: Routledge, 2006.

Chun, Wendy Hui Kyong. *Programmed Visions: Software and Memory*. Cambridge, MA: MIT Press, 2011.

Coates Nee, Rebecca, and Mariana De Maio. "A 'Presidential Look'? An Analysis of Gender Framing in 2016 Persuasive Memes of Hillary Clinton." *Journal of Broadcasting & Electronic Media* 63, no. 2 (2019): 304–321.

Cohan, John Alan. "'I Didn't Know' and 'I Was Only Doing My Job': Has Corporate Governance Careened Out of Control? A Case Study of Enron's Information Myopia." *Journal of Business Ethics* 40, no. 3 (2002): 275–299.

Cohen, William W. "Enron Email Dataset." Carnegie Mellon University, 2015. https://www.cs.cmu.edu/~./enron/.

Cole, David. "We Kill People Based on Metadata." *New York Review of Books*, May 10, 2014. http://www.nybooks.com/daily/2014/05/10/we-kill-people-based-metadata/.

"comScore Releases First Comparative Report on Mobile Usage in Japan, United States and Europe." comScore, October, 2010. http://www.comscore.com/Insights/Press-Releases/2010/10/comScore-Release-First-Comparative-Report-on-Mobile-Usage-in-Japan-United-States-and-Europe.

Conrad, Rita-Marie. 1999. "Save Yourself from Drowning in Online Interaction." Designing for Learning, January 1999.

Cook, Elizabeth Heckendorn. *Epistolary Bodies: Gender and Genre in the Eighteenth-Century Republic of Letters*. Stanford, CA: Stanford University Press, 1996.

Coopersmith, Jonathan. "Gawker Removing History." SIGCIS, November 6, 2016. http://lists.sigcis.org/pipermail/members-sigcis.org/2016-November/000364.html.

Couldry, Nick, Sonia Livingstone, and Tim Markham. *Media, Society, World: Social Theory and Digital Media Practice*. Cambridge: Polity, 2013.

Couldry, Nick and Andreas Hepp. *The Mediated Construction of Reality*. Cambridge: Polity, 2017.

Couldry, Nick, Sonia Livingstone, and Tim Markham. "'Public Connection' and the Uncertain Norms of Media Consumption." In *Citizenship and Consumption*, edited by Kate Soper and Frank Trentmann. Basingstoke: Palgrave Macmillan, 2007.

Crawford, Kate. "Following You: Disciplines of Listening in Social Media." *Continuum* 23, no. 4 (2009): 525–535. https://doi.org/10.1080/10304310903003270.

Crocker, Dave. "Re: V.A. Shiva Ayyadurai: Inventor of E-mail Honored by Smithsonian." *IETF*, February 18, 2012. https://www.ietf.org/mail-archive/web/ietf-smtp/current/msg00148.html.

Crocker, Dave. "A History of E-mail: Collaboration, Innovation and the birth of a System." *Washington Post*, March 20, 2012. https://www.washingtonpost.com/national/on-innovations/a-history-of-e-mail-collaboration-innovation-and-the-birth-of-a-system/2012/03/19/gIQAOeFEPS_story.html.

Crocker, Stephen. Documentation Conventions. RFC 3, https://doi.org/10. 17487/RFC0003, April 1969.

Crystal, David. *Internet Linguistics: A Student Guide*. London: Routledge, 2011.

Cyger, Michael. "After 15 Years, Fastmail Finally Acquires Their .Com—With Rod Mueller." *Domain Sherpa*, November 17, 2014.

Danet, Brenda. *Cyberpl@y: Communicating Online*. London: Bloomsbury, 2001.

Davis, Allison P. "Two French Unions Ban Checking Work Emails after 6 p.m." *NYMAG.com*, April 10, 2014.

"Declaration of John F. Hackett." *Jason Leopold v. US Department of State*. Case 1:15-cv-00123-RC, filed May 18, 2015 (US District Court for the District of Columbia, 2015).

Deitrick, William, Zachary Miller, Benjamin Valyou, and Wei Hu. "Author Gender Prediction in an Email Stream Using Neural Networks." *Journal of Intelligent Learning Systems and Applications* 4, no. 3 (2012): 169–175.

DeNardis, Laura. *Protocol Politics: The Globalization of Internet Governance*. Cambridge, MA: MIT Press, 2009.

Denton, Nick. "A Hard Peace." *Nick Denton Blog*, November 2, 2016. https://nickdenton.org/a-hard-peace-e161e19bfaf#.orhanu220.

Diesner, Jana, Terrill Frantz, and Kathleen Carley. "Communication Networks from the Enron Email Corpus: 'It's Always About the People. Enron is no Different,'" *Computational & Mathematical Organization Theory* 11 (2005): 201–228.

Dinan, Stephen. "Hillary Clinton Deleted 32,000 Private Emails, Refuses to Turn Over Server." *Washington Times*, March 10, 2015. http://www.washingtontimes.com/news/2015/mar/10/hillary-clinton-deleted-32000-private-emails-refus/.

Dinan, Stephen. "State Department Can't Find Enough Staffers to Process Hillary Clinton's Emails." *Washington Times*, October 13, 2015. http://www.washingtontimes.com/news/2015/oct/13/hillary-clinton-emails-state-department-cant-find-/.

Ding, Wei, Peng Liang, Antony Tang, and Hans Van Vliet. "Understanding the Causes of Architecture Changes using OSS Mailing Lists." *International Journal of Software Engineering and Knowledge Engineering* 25, no. 9 (2015): 1633–1651.

"Diversify My Emoji." Accessed April 9, 2017. https://www.dosomething.org/us/campaigns/diversify-my-emoji.

Douglas, Susan J, "Some Thoughts on the Question 'How Do New Things Happen?'" *Technology and Culture* 51, no. 2 (April 2010): 293–304.

Dover, Mike. "More News About the Death of Email." *Wikinomics*, June 18, 2008. http://www.wikinomics.com/blog/index.php/2008/06/18/more-news-about-the-death-of-email/.

Du Gay, Paul. 2000. *In Praise of Bureaucracy: Weber, Organization, Ethics*. London: Sage, 2000.

Dubin, Fraida. "Checking out E-mail and the Fax." *English Today* 7, no. 1 (1991): 47–51.

Duckett, Chris. 2013. "FastMail Staff Buy Back Company from Opera." *ZDNet*, September 26, 2013.

Duffy, Brooke Erin, and Emily Hund. "'Having It All' on Social Media: Entrepreneurial Femininity and Self-Branding Among Fashion Bloggers." *Social Media + Society* 1, no. 2 (2014): 1–11.

Echols, Micah. 2003. "Striking a Balance Between Employer Business Interests and Employee Privacy: Using *Respondeat Superior* to Justify the Monitoring of Web-Based, Personal Electronic Mail Accounts of Employees in the Workplace." *Computer Law Review and Technology Journal* 7 (2003): 273–300.

ECOLOG-L@LISTSERV.UMD.EDU. Ecological Society of America, accessed May 28, 2017. http://www.lsoft.com/scripts/wl.exe?SL1=ECOLOG-L&H=LISTSERV.UMD.EDU.

"Email Client Market Share." Litmus Email Analytics. http://emailclientmarketshare.com/.

Radicati Group, The. "Email Statistics Report, 2020–2024." March 2, 2020. https://www.radicati.com/?p=16510.

emailmonday. "The Ultimate Mobile Email Statistics Overview." Accessed April 9, 2017. http://www.emailmonday.com/mobile-email-usage-statistics.

"EMMS; Electronic Mail and Message Systems; Electronic Mail and Micro Systems; Electronic Mail and Messaging Systems," The Computer History Museum, Catalog Number 102661013, 1977–2001. http://www.computerhistory.org/collections/catalog/102661013.

"Enron's Collapse; Excerpts From a House Hearing on Destruction of Enron Documents." *New York Times*, January 25, 2002. http://www.nytimes.com/2002/01/25/business/25FULL-TEXT-HOUSE.html.

Enron Email Data. Amazon Web Services. Accessed June 11, 2017. https://aws.amazon.com/datasets/enron-email-data/.

"Enron Fast Facts." *CNN*. Accessed April 27, 2017. http://edition.cnn.com/2013/07/02/us/enron-fast-facts/.

Environmental Paper Network. 2008. "Increasing Paper Efficiency."

"Exhibition Listings." Preview. Accessed June 11, 2017. http://www.preview-art.com/previews/02-2010/mailart.html.

Faba-Pérez, Cristina, and Ana-María Cordero-González. "The Validity of Bradford's Law in Academic Electronic Mailing Lists." *The Electronic Library* 33, no. 6 (2015): 1043–1044.

Farber, David, and Paul Baran. "The Convergence of Computing and Telecommunications Systems." *Science* 195, no. 4283 (March 18, 1977): 1166–1170. https://doi.org/10.1126/science.195.4283.1166.

Farivar, Cyrus. "Judge Dismisses Shiva 'I Invented EMAIL' Ayyadurai's Libel Lawsuit against Techdirt." *Ars Technica*, September, 7, 2017. https://arstechnica.com/tech-policy/2017/09/judge-dismisses-libel-lawsuit-filed-by-self-proclaimed-e-mail-inventor/.

Fasbinder, Joe. "Virtual Love and Cyber-Romance." *UPI*, February 12, 1995. http://www.upi.com/Archives/1995/02/12/Virtual-love-and-cyber-romance/9702792565200/.

Favret, Mary A. *Romantic Correspondence: Women, Politics and the Fiction of Letters*. Cambridge: Cambridge University Press, 2007.

Fazekas, David. "Hillary Clinton's 55,000 Pages of Emails: A Whole Lot of Paper." *ABC News*, July 5, 2016. http://abcnews.go.com/Politics/hillary-clintons-55000-pages-emails-lot-paper/story?id=32663275.

Federal Trade Commission. "Spam." http://www.consumer.ftc.gov/articles/0038-spam.

Feifer, Jason. "How 'Structure' Saved Mailchimp." *Entrepreneur* 46, no. 6 (2018): 31.

Feinler, Elizabeth. "Who Invented Email?" *SIGCIS*, February 29, 2012. http://www.sigcis.org/.

Ferguson, Kathy. *The Feminist Case Against Bureaucracy*. Philadelphia: Temple University Press, 1984.

Flanagan, Heather. "The RFC Series and the Twenty-first Century." *IETF Journal*, July 1, 2014.

Flanagan, Patrick. "Spam: Not Just for Breakfast Anymore: Unsolicited Email in the Business Environment." 4th Annual Ethics and Technology Conference, June 4–5, 1999.

Fletcher, Stephanie. *E-Mail: A Love Story*. London: Headline Book Publishing, 1996.

Fourteenth Annual Report of the Article 29 Working Party on Data Protection. Luxembourg: Publications Office of the European Union, 2013.

Fox News. "State Department Releases Over 3,000 Clinton Emails on New Year's Eve." January 1, 2016. http://www.foxnews.com/politics/2016/01/01/state-department-releases-over-3000-clinton-emails-on-new-years-eve.html.

Fox, Susannah, and Lee Rainie. "The Web at 25." Pew Research Center, February 25, 2014. http://www.pewinternet.org/2014/02/25/the-web-at-25-in-the-u-s.

Fran Martin, ed. *Interpreting Everyday Culture*. London: Hodder Arnold, 2003.

Frehner, Carmen. *Email—SMS—MMS: The Linguistic Creativity of Asynchronous Discourse in the New Media Age*. New York: Linguistic Insights, 2008.

French Tech Workers: No E-Mails After 6. *FOXBusiness*, April 10, 2014. http://www.foxbusiness.com/personal-finance/2014/04/10/french-tech-workers-no-e-mails-after-6/.

Fuata, Brian. "All Titles, No Centre Crux: The Email Performances in PERFORMA 2015." *PERFORMA 15*.

Fuller, Matthew, and Andrew Goffey. *Evil Media*. Cambridge, MA: MIT Press, 2012.

Furlong, Jordan. "The Last Days of E-mail." *Law 21* (blog), February 29, 2008.

Galloway, Alexander R. *The Interface Effect*. Cambridge: Polity, 2012.

Galloway, Alexander R. *Protocol: How Control Exists after Decentralization*. Cambridge, MA: MIT Press, 2004.

Garling, Caleb. "Who Invented Email? Just Ask…Noam Chomsky." *Wired*, June 16, 2012.

Garton, Laura, and Barry Wellman, "Social Impacts of Electronic Mail in Organisations: A Review of the Research Literature," *Communication Yearbook* 18 (1995): 434–453.

Gershon, Ilana. *The Breakup 2.0: Disconnecting over New Media*. Ithaca, NY: Cornell University Press, 2010.

Gilbert, Eric. "Phrases That Signal Workplace Hierarchy." Presented at the ACM Conference on Computer Supported Cooperative Work, Seattle, WA, USA, 2012.

Gill, Rosalind, and Andy Pratt. "In the Social Factory? Immaterial Labour, Precariousness and Cultural Work." *Theory, Culture & Society* 25, no. 7–8 (2008): 1–30.

Gilroy, Amanda, and W. M. Verhoeven. *Epistolary Histories: Letters, Fiction, Culture.* Charlottesville: University Press of Virginia, 2000.

Gitelman, Lisa. *Paper Knowledge: Toward a Media History of Documents*. Durham, NC: Duke University Press, 2014.

Gleeson, Jessamy. "(Not) 'Working 9–5': The Consequences of Contemporary Australian-Based Online Feminist Campaigns as Digital Labour." *Media International Australia* 161, no. 1 (2016).

Goddard, Michael. "Opening Up the Black Boxes: Media Archaeology, 'Anarchaeology' and Media Materiality." *New Media & Society* 17 (2015): 1761–1776.

Goggin, Gerard, and Larissa Hjorth. "Introduction: Mobile Media Research—State of the Art." In *The Routledge Companion to Mobile Media*, edited by Gerard Goggin and Larissa Hjorth. New York: Routledge, 2013.

Goggin, Gerard, and Mark McLelland. *The Routledge Companion to Global Internet Histories*. London: Routledge, 2017.

Golden, Catherine. 2010. *Posting It: The Victorian Revolution in Letter Writing*. Gainesville: University Press of Florida.

Goldsmith, Elizabeth C., ed. *Writing the Female Voice*. Boston: Northeastern University Press, 1989.

"Good Technology Survey Reveals Americans Working More, But On Their Own Schedule." *PR Newswire*, July 2, 2012. http://www.prnewswire.com/news-releases/good-technology-survey-reveals-americans-working-more-but-on-their-own-schedule-161018305.html.

Goode, Joanne, and Maggie Johnson. "Putting out the Flames: The Etiquette and Law of E-Mail." *ONLINE: The Magazine of Online Information Systems* 15, no. 6 (1991): 61–65.

Goodsell, Charles T. *The New Case for Bureaucracy*. Los Angeles: Sage/CQ Press, 2015.

Grau, Oliver. *Virtual Art: From Illusion to Immersion*. Cambridge, MA: MIT Press, 2003.

Greenberg, Andy. "How Reporters Pulled Off the Panama Papers, The Biggest Leak in Whistleblower History." *Wired*, April 4, 2016.

Greenfield, Rebecca. "The Other Tech Figure Who's Trying to Kill Gawker." *Bloomberg*, June 4, 2016. https://www.bloomberg.com/news/articles/2016-06-03/the-other-tech-figure-who-s-trying-to-kill-gawker.

Gregg, Melissa. *Counterproductive: Time Management in the Knowledge Economy*. Durham, NC: Duke University Press, 2018.

Gregg, Melissa. "The Normalisation of Flexible Female Labour in the Information Economy." *Feminist Media Studies* 8, no. 3 (2008): 285–299.

Gregg, Melissa. *Work's Intimacy*. Cambridge: Polity, 2011.

Guevara, Marina Walker. 2016. "ICIJ Releases Database Revealing Thousands of Secret Offshore Companies." *The International Consortium of Investigative Journalists*, May 9, 2016. https://panamapapers.icij.org/blog/20160509-offshore-database-release.html.

Gutwin, Karl, Reagan Penner, and Kevin Schneider. "Group Awareness in Distributed Software Development." Presented at the ACM Conference on Computer Supported Cooperative Work, Chicago, IL, USA, 2004.

Guzzi, Anja, Alberto Bacchelli, Michele Lanza, Martin Pinzger, and Arie van Deursen. "Communication in Open Source Software Development Mailing Lists." 10th Working Conference on Mining Software Repositories, San Francisco, California, 2013.

Hafner, Katie, and Matthew Lyon. *Where Wizards Stay Up Late*. New York: Simon & Schuster, 1996.

Haag, Steffi, Andreas Eckhardt, and Andrew Schwarz, "The Acceptance of Justifications among Shadow IT Users and Nonusers—An Empirical Analysis." *Information & Management* 56, no. 5 (July 2019): 731–741.

Haigh, Thomas. "Did V.A. Shiva Ayyadurai Invent Email?" *SIGCIS*, August 4, 2015. http://www.sigcis.org/ayyadurai.

Haigh, Thomas. "Email was Invented by a School Boy in 1978 Says Washington Post & Time Magazine." *SIGCIS*, February 22, 2012. http://www.sigcis.org/.

Haigh, Thomas. "Protocols for Profit: Web and E-mail Technologies as Product and Infrastructure." In *The Internet and American Business*, edited by William Aspray and Paul E Ceruzzi. Cambridge, MA: MIT Press, 2008.

Hall, Stuart. "Cultural Studies and Its Theoretical Legacies." In *Cultural Studies*, edited by Cary Nelson, Lawrence Grossberg, and Paula Treichler. Routledge, 1992.

Hancock, Philip, and Melissa Tyler. "'MOT Your Life': Critical Management Studies and the Management of Everyday Life." *Human Relations* 55, no. 5 (2004): 619–645.

Haraway, Donna *Modest_Witness@Second_Millennium. FemaleMan©_Meets_OncoMouse TM: Feminism and Technoscience*. New York: Routledge, 1997.

Haraway, Donna. "Situated Knowledges: The Science Question in Feminism and the Privilege of Partial Perspective." *Feminist Studies* 14, no. 3 (1988): 575–599.

Hardin, Johanna, and Ghassan Sarkis. "Network Analysis with the Enron Email Corpus." *Journal of Statistics Education* 23, no. 2 (2015).

Harman, Donna. "It's Better for the Environment to Send a Letter." *New York Times*, September 23, 2012.

Hauben, Michael, and Ronda Hauben. *Netizens: On the History and Impact of Usenet and the Internet*. Los Alamitos, CA: Computer Science Press, 1997.

Hayward, John. "Laws Are for the Little People: State Dept. Violates Court Order on Clinton Emails." *Breitbart*, December 31, 2015. http://www.breitbart.com/big-government/2015/12/31/laws-little-people-state-dept-violates-court-order-clinton-emails/.

Held, John. 2000. "Living Thing in Flight: Contributions and Liabilities of Collecting and Preparing Contemporary Avant-Garde Materials for an Archive." *Archives of American Art Journal* 40, no. 3/4 (2000).

Held, John. "The Mail Art Exhibition: Personal Worlds to Cultural Strategies." In *At a Distance: Precursors to Art and Activism on the Internet*, edited by Annmarie Chandler and Norie Neumark. Cambridge, MA: MIT Press, 2005.

Held, John. *Small Scale Subversion: Mail Art & Artistamps*. Breda, Netherlands: TAM Publishing, 2015.

Helman, Igor. "Spam-a-Lot: The States' Crusade Against Unsolicited Email in Light of the CAN-SPAM Act and the Overbreadth Doctrine." *Boston College Law Review* 50, no. 5 (2009).

Henderson, Austin. "Re: Contents of Subject Fields." *MSGGRP*, April 25, 1977.

Herath, Tejaswini, Rui Chen, Jingguo Wang, Ketan Banjara, Jeff Wilbur, and H. Raghav Rao. "Security Services as Coping Mechanisms: An Investigation into User Intention to Adopt an Email Authentication Service." *Information Systems Journal* 24, no. 1 (January 2014): 61–84.

Herzfeld, Michael. *The Social Production of Indifference: Exploring the Symbolic Roots of Western Bureaucracy*. Chicago: The University of Chicago Press, 1992.

Hicks, Mar. *Programmed Inequality How Britain Discarded Women Technologists and Lost Its Edge in Computing*. Cambridge, MA: MIT Press, 2017.

Highfield, Tim, and Axel Bruns. "Confrontation and Cooptation: A brief History of Australian Political Blogs." *Media International Australia* 143 (2012): 89–98.

Highfield, Tim, and Tama Leaver. "Instagrammatics and Digital Methods: Studying Visual Social Media, from Selfies and GIFs to Memes and Emoji." *Communication Research and Practice* 2, no. 1 (2016): 47–62.

Highmore, Ben. 2002. *Everyday Life and Cultural Theory*. New York: Taylor and Francis, 2002.

Highmore, Ben. 2017. "Georges Perec and the Significance of the Insignificant." In *The Afterlives of Georges Perec*, edited by Rowan Wilken and Justin Clemens. Edinburgh: Edinburgh University Press.

Hidalgo, César, interviewed by Abraham Riesman. "What Your Metadata Says about You." *Boston Globe*, June 30, 2013.

"Hillary Clinton's Benghazi Emails—Live Dive. *Wall Street Journal*, May 22, 2015. http://blogs.wsj.com/washwire/2015/05/22/hillary-clintons-benghazi-emails-live-dive.

Hilton, Chad, and Naoki Kameda. "E-Mail and the Internet as International Business Communication Teaching and Research Tools—A Case Study." *Journal of Education for Business* 74, no 3 (1999): 181–186.

Hjorth, Larissa. *Mobile Media in the Asia-Pacific: Gender and the Art of Being Mobile*. London: Routledge, 2011.

Hochschild, Arlie R. *The Managed Heart: Commercialization of Human Feeling*. Berkeley: University of California Press, 2003.

Hodson, Thomas J., Fred Englander, and Valerie Englander. "Ethical, Legal and Economic Aspects of Employer Monitoring of Employee Electronic Mail." *Journal of Business Ethics* 19 (1999): 99–108.

Holmes, Ryan. "Email Is the New Pony Express—And It's Time to Put It Down." *Fast Company*, October 16, 2012.

"How Are Mobile, Desktop and Webmail Categories Defined." Litmus, n.d. https://litmus.com/help/analytics/category-definitions/.

"How Gmail's New Inbox Is Affecting Open Rates." *Mailchimp* (blog), July 23, 2013. https://web.archive.org/web/20130828150415/http://blog.mailchimp.com/how-gmails-new-inbox-is-affecting-open-rates/.

"How We Set Up Emoji Support for Subject Lines." *Mailchimp* (blog), February 25, 2015. https://mailchimp.com/resources/how-we-set-up-emoji-support-for-subject-lines/.

Huhtamo, Erkki. "Dismantling the Fairy Engine Media Archaeology as Topos Study." In *Media Archaeology Approaches, Applications, and Implications*, edited by Erkki Huhtamo and Jussi Parikka. Berkeley: University of California Press, 2011.

Huhtamo, Erkki, and Jussi Parikka. "Introduction: An Archaeology of Media Archaeology." In *Media Archaeology Approaches, Applications, and Implications*, edited by Erkki Huhtamo and Jussi Parikka. Berkeley: University of California Press, 2011.

Hummel, Ralph P. *The Bureaucratic Experience: The Post-modern Challenge*. Armonk, NY: M. E. Sharpe, 2008.

IBM. "2016 Email Marketing Metrics Benchmark Study." accessed April 9, 2017. http://www.silverpop.com/marketing-resources/white-papers/all/2016/email-metrics-benchmark-study-2016/.

IEEE Computer Society. "IEEE Internet Award Recipients." https://www.ieee.org/about/awards/bios/internet_recipients.html.

"Inductees: Internet Hall of Fame Innovator, Raymond Tomlinson." Internet Hall of Fame. Accessed June 23, 2017. http://internethalloffame.org/inductees/raymond-tomlinson.

Ingold, Tim. *Being Alive: Essays on Movement, Knowledge and Description*. New York: Routledge, 2011.

Ishii, Kenichi. "Implications of Mobility: The Uses of Personal Communication Media in Everyday Life." *Journal of Communication* 56 (2006): 346–365.

Ito, Mizuko, and Daisuke Okabe. "Technosocial Situations: Emergent Structuring of Mobile E-mail Use." In *Personal, Portable Intimate: Mobile Phones in Japanese Life*, edited by Misa Matsuda, Mizuko Ito, and Daisuke Okabe. Cambridge, MA: MIT Press, 2005.

Ito, Mizuko, Sonja Baumer, Matteo Bittanti, danah boyd, Rachel Cody, Becky Herr Stephenson, Heather A. Horst, Patricia G. Lange, Dilan Mahendran, Katynka Z. Martínez, C. J. Pascoe, Dan Perkel, Laura Robinson, Christo Sims, and Lisa Tripp. *Hanging Out, Messing Around, and Geeking Out: Kids Living and Learning with New Media*. Cambridge, MA: MIT Press, 2010.

Jakobs, Kai. "Why Then Did the X.400 E-mail Standard Fail? Reasons and Lessons to be Learned." *Journal of Information Technology* 28, no. 1 (2013): 63–73.

Jarrett, Kylie. *Feminism, Labour and Digital Media: The Digital Housewife*. New York: Routledge, 2016.

John, Jeremy Leighton. "Digital Forensics and Preservation." Digital Preservation Coalition in association with Charles Beagrie Ltd, November 2012. http://www.dpconline.org/docman/technology-watch-reports/810-dpctw12-03-pdf/file.

John, Nicholas A., and Asaf Nissenbaum. "An Agnotological Analysis of APIs: Or, Disconnectivity and the Ideological Limits of Our Knowledge of Social Media." *The Information Society* 35, no. 1 (2019): 1–12, https://doi.org/10.1080/01972243.2018.1542647.

Johns, Amelia, and Abbas Rattani. "Somewhere in America": The #MIPSTERZ Digital Community and Muslim Youth Voice Online." In *Negotiating Digital Citizenship: Control, Contest and Culture*, edited by Sonja Vivienne, Amellia Johns, and Anthony McCosker. London: Rowman & Littlefield, 2016.

Johnson, Eugene C. Testimony, United States Postal Service Recommended Decision on Changes in the Classification Schedule (ECOM), Docket Number MC78–3, Postal Rate Commission, September 8, 1978, 8.

Johnson, John. 2014. "France Bans Work E-mail After 6 p.m." *USA Today*, April 11, 2014. http://www.usatoday.com/story/money/business/2014/04/11/newser-france-work-email-ban/7592125/.

Jones, Alasdair. "Everyday Without Exception? Making Space for the Exceptional in Contemporary Sociological Studies of Streetlife." *The Sociological Review* 66, no. 5 (2018): 1000–1016.

Jordan, Justine. "Top 10 Most Popular Email Clients of 2015." *Litmus* (blog), December 15, 2015. https://litmus.com/blog/top-10-most-popular-email-clients-of-2015.

Julienne, Hanna. "World Geopolitics Through Hillary Clinton's Emails." *Dataiku*, October 25, 2015.

kaggle. "Hillary Clinton's Emails: Uncover the Political Landscape in Hillary Clinton's Emails." *kaggle* (blog), June 11, 2017. https://www.kaggle.com/kaggle/hillary-clinton-emails.

Kafka, Ben. 2012. *The Demon of Writing: Powers and Failures of Paperwork*. Brooklyn, NY: Zone Books.

Kakuko, Miyata, Jeffrey Boase, and Barry Wellman. "The Social Effects of Keitai and Personal Computer E-mail in Japan." In *Handbook of Mobile Communication Studies*, edited by James E. Katz. Cambridge MA: MIT Press, 2006.

Kaneshige, Tom. "Hillary Clinton is Now the Face of Shadow IT." *CIO*, March 12, 2015.

Kang, Hyunmo, Catherine Plaisant, Tamer Elsayed, and Douglas Oard. "Making Sense of Archived E-mail: Exploring the Enron Collection With NetLens." *Journal of the American Society for Information Science and Technology* 61, no 4 (2009): 723–744.

Kauffman, Linda. *Discourses of Desire: Gender, Genre, and Epistolary Fictions*. Ithaca: Cornell University Press, 1986.

Kauffman, Linda. *Special Delivery: Epistolary Modes in Modern Fiction*. Chicago: University of Chicago Press, 1992.

Kazienko, Przemysław, Katarzyna Musiał, and Aleksander Zgrzywa. "Evaluation of Node Position Based on Email Communication." *Control and Cybernetics* 38, no. 1 (2009): 67–86.

Keaten, James, and Lynne Kelly. "Re: We Really Need to Talk': Affect for Communication Channels, Competence, and Fear of Negative Evaluation." *Communication Quarterly* 56, no. 4 (2008): 407–426.

Kelner, Simon. "French Workers Now Have Legal Right Not to be Contacted After They Leave the Office." *The Independent*, April 9, 2014. http://www.independent.co.uk/news/world/europe/french-workers-now-have-legal-right-not-to-be-contacted-after-they-leave-the-office-9249506.html.

Kennedy, Jenny, and Esther Milne. "Public Privacy: Reciprocity and Silence." *Platform: Journal of Media and Communication* 4, no. 2 (2013).

Kiely, Eugene. "More Spin on Clinton Emails." FactCheck.org, September 2015. http://www.factcheck.org/2015/09/more-spin-on-clinton-emails/.

Kiely, Eugene. "Clinton's Email Narrative, Interrupted." FactCheck.org, September 2015. http://www.factcheck.org/2015/09/clintons-email-narrative-interrupted/.

Kigerl, Alex C. "Deterring Spammers: Impact Assessment of the CAN SPAM Act on Email Spam Rates." *Criminal Justice Policy Review* 27, no. 8 (2016): 791–811.

Kim, Kyoung-hwa Yonnie. "Genealogy of Mobile Creativity: A Media Archaeological Approach to Literary Practice in Japan." In *The Routledge Companion to Mobile Media*, edited by Gerard Goggin and Larissa Hjorth. New York: Routledge, 2013.

Kininmonth, Sam. "Benevolent Dictatorships? Moderation as a Tool of Platform Governance on Reddit." Honors Thesis. Swinburne University, 2016.

Kirchner, Lauren. "AOL Settled with Unpaid 'Volunteers' for $15 Million." *Columbia Journalism Review*, February 10, 2011.

Kirschenbaum, Matthew G., Richard Ovenden, and Gabriela Redwine. Digital *Forensics and Born-Digital Content in Cultural Heritage Collections*. Washington, DC: Council on Library and Information Resources, 2012.

Kittler, Frederich. *Gramophone, Film, Typewriter*. Stanford, CA: Stanford University Press, 1999.

Klensin, John. 2001. Simple Mail Transfer Protocol. RFC 2821, https://doi.org/10.17487/RFC2821, April 2001.

Klensin, John. 2008. Simple Mail Transfer Protocol. RFC 5321, https://doi.org/10.17487/RFC5321, October 2008.

Klimt, Bryan, and Yiming Yang. "The Enron Corpus: A New Dataset for Email Classification Research." In *Lecture Notes in Computer Science*, vol. 3201. Berlin: Springer, 2004.

Kluitenberg, Eric. "On the Archaeology of Imaginary Media." In *Media Archaeology Approaches, Applications, and Implications*, edited by Erkki Huhtamo and Jussi Parikka. Berkeley: University of California Press, 2011.

Knibbs, Kate. "11 of the Most Obnoxious Email Signatures Ever." *Gizmodo*, March 3, 2015.

Knutsen, Lars A., and Kalle Lyytinen. "Messaging Specifications, Properties and Gratifications as Institutions: How Messaging Institutions Shaped Wireless Service Diffusion in Norway and Japan." *Information and Organization* 18 (2008).

Kolawole, Emi. "Smithsonian Acquires Documents from Inventor of 'EMAIL' Program." *Washington Post*, February 17, 2012. https://www.washingtonpost.com/national/on-innovations/va-shivaayyadurai-inventor-of-e-mail-honored-by-smithsonian/2012/02/17/gIQA8gQhKR_story.html.

Korte, Gregory, and Edward C. Baig. "Could Clinton's BlackBerry Handle Multiple E-mail Accounts?" *USA Today*, March 10, 2015. http://www.usatoday.com/story/news/politics/2015/03/10/hillary-clinton-emails-blackberry/24725993/.

Kushin, Matthew, Masahiro Yamamoto, and Francis Dalisay, "Societal Majority, Facebook, and the Spiral of Silence in the 2016 US Presidential Election." *Social Media + Society* (April–June 2019): 1–11.

Kuus, Merje. "Transnational Bureaucracies: How Do We Know What They Know?" *Progress in Human Geography* 39, no. 4 (2015): 432–448.

Lazzarato, Maurizio. "Immaterial Labor." In *Radical Thought in Italy: A Potential Politics*, edited by P. Virno and M. Hardt. Minneapolis: University of Minnesota Press, 1996.

Lea, Martin, et al. "'Flaming' in Computer-Mediated Communication: Observations, Explanations, Implications." In *Contexts of Computer-Mediated Communication*, edited by Martin Lea. Hertfordshire: Harvester, 1992.

Leber, Jessica. "The Immortal Life of the Enron E-mails." *MIT Technology Review*, July 2, 2013.

Lee, M. J. "Hillary Clinton Donors Frustrated by Email Controversy." *CNN Politics*, March 10, 2015. http://edition.cnn.com/2015/03/10/politics/donors-hillary-clinton-emails.

Lehnert, Wendy G. *Internet 101: A Beginner's Guide to the Internet and the World Wide Web*. Reading, MA: Addison Wesley Longman, 1998.

Leonard, Devin. *Neither Snow Nor Rain: A History of the United States Postal Service*. New York: Grove Press, 2016.

Letter from National Archives, Chief Records Officer, Paul M. Wester to Deputy Assistant Secretary for Global Information Services, Margaret P. Grafield. 2015. "Possible Alienation of Former Secretary of State, Hillary Clinton's Federal Email Records." National Archives.

"Letter to Hillary Clinton's Representative, Cheryl Mills re: the Federal Records Act of 1950, November 12. 2014." National Archives.

Lewkowicz, Kayla. "Harness the Power of Emojis in Your Inbox." *Litmus* (blog), February 13, 2017. https://litmus.com/blog/harness-the-power-of-emojis-in-your-inbox.

Lewcovicz, Kayla. "Webmail Increases to 29% for November Email Client Market Share." *Litmus* (blog), December 7, 2016. https://litmus.com/blog/webmail-increases-to-29-for-november-email-client-market-share.

Leys, Ruth. "The Turn to Affect: A Critique." *Critical Inquiry* 37, no. 3 (2011): 434–472.

Lifshitz, Lisa R. "CASL—How To Send E-mails to Canadians Safely" *Business Law Today*, April 4, 2016. http://www.americanbar.org/publications/blt/2016/04/04_lifshitz.html.

Ling, Richard. *Taken for Grantedness: The Embedding of Mobile Communication into Society*. Cambridge, MA: MIT Press, 2012.

Litmus. "Which Gmail Tab Will Your Email Appear Under?" Accessed July 25, 2017. https://litmus.com/gmail-tabs.

Livingstone, Sonia. "On the Material and the Symbolic: Silverstone's Double Articulation of Research Traditions in New Media Studies." *New Media & Society* 9, no. 1 (2007).

Lobato, Ramon. *Shadow Economies of Cinema: Mapping Informal Film Distribution*. London: British Film Institute/Palgrave Macmillan, 2012.

"Lomholt Mail Art Archive." Accessed June 11, 2017. http://www.lomholtmail artarchive.dk/intro.

Lorenz, Chad. "The Death of E-Mail: Teenagers are Abandoning their Yahoo! and Hotmail Accounts. Do the Rest of us Have To?" *Slate*, November 14, 2007.

Lucas, Gavin. *The Story Of Emoji*. Munich: Prestel Publishing, 2016.

Luhmann, Niklas. *Social Systems*. Translated by John Bednarz Jr. and Dirk Baecker. Stanford, CA: Stanford University Press, 1995.

Mabry, Edward A. "Framing Flames: The Structure of Argumentative Messages on the Net." *Journal of Computer-Mediated Communication* 2, no. 4, (1997).

Madianou, Mirca, and Daniel Miller. "Polymedia: Towards a New Theory of Digital Media in Interpersonal Communication." *International Journal of Cultural Studies* 16, no. 2 (2013): 169–187.

Mailchimp. "About Mailchimp." Accessed April 9, 2017. https://mailchimp.com/about.

Mailchimp. "About Gmail Tabs." Accessed March 2, 2017. https://mailchimp.com /help/about-gmail-tabs.

Mailchimp. "Common Rookie Mistakes for Email Marketers." Accessed April 13, 2020. https://mailchimp.com/resources/common-rookie-mistakes-email-marketers/.

"Mailchimp's Most Popular Subject Line Emojis." *Mailchimp* (blog), May 4, 2015. https://mailchimp.com/resources/mailchimps-most-popular-subject-line-emojis/.

Main, Lisa. "Data Retention: What Is Metadata and How Will It Be Defined by New Australian Laws?" *ABC News*, March 17, 2015. http://www.abc.net.au/news/2015-03 -17/what-is-metadata-how-will-it-be-defined-by-new-australia-laws/6325908.

Mangan, Lucy. "When the French Clock Off at 6pm, They Really Mean It." *Guardian*, April 9, 2014. http://www.theguardian.com/money/shortcuts/2014/apr/09 /french-6pm-labour-agreement-work-emails-out-of-office.

Marshall, Jonathan Paul. *Living on Cybermind: Categories, Communication, and Control* New York: Peter Lang, 2007.

Masnick, Mike. "The Latest on Shiva Ayyadurai's Failed Libel Suit against Techdirt". *Techdirt,* October 6, 2017. https://www.techdirt.com/articles/20171006/11584638359/latest-shiva-ayyadurais-failed-libel-suit-against-techdirt.shtml.

Mathis-Lilley, Ben. "Hillary Clinton's 'Thorough' Email Review Was a Keyword Search." *Slate,* March 13, 2015.

Matias, J. Nathan. "The Civic Labor of Online Moderators." Paper presented at The Internet Politics and Policy Conference, Oxford, 2016.

Matzat, Uwe. "Quality of Information in Academic E-mailing Lists." *Journal of the American Society for Information Science and Technology* 60, no. 9 (2009): 1859–1870.

Matzat, Uwe. "Disciplinary Differences in the Use of Internet Discussion Groups: Differential Communication Needs or Trust Problems?" *Journal of Information Science* 35, no. 5 (2009): 613–631.

Mazmanian, Melissa, Wanda Orlikowski, and Joanne Yates, "The Autonomy Paradox: The Implications of Mobile Email Devices for Knowledge Professionals." *Organization Science* 24, no. 5 (2013): 1337–1357.

McCallum, Andrew, Xuerui Wang, and Andres Corrada-Emmanuel. "Topic and Role Discovery in Social Networks with Experiments on Enron and Academic Email." *Journal of Artificial Intelligence Research* 30 (2007): 249–272.

McCarthy, John. "Networks Considered Harmful for Electronic Mail." *Communications of the ACM* 32, no. 12 (1989): 1389–1390.

McCarthy, Nan. *Chat: A Cybernovel.* Berkeley, CA: Peachpit Press, 1996.

McCormack, Tom. "Emoticon, Emoji, Text II: Just ASCII." *Rhizome* (blog), April 30, 2013. http://rhizome.org/editorial/2013/apr/30/emoticon-emoji-text-ii-ascii.

McCosker, Anthony. *Intensive Media: Aversive Affect and Visual Culture.* Basingstoke: Palgrave Macmillan, 2013.

McCosker, Anthony, and Rowan Wilken. "'Things That Should Be Short': Perec, Sei Shōnagon, Twitter, and the Uses of Banality." In *The Afterlives of Georges Perec,* edited by Rowan Wilken and Justin Clemens. Edinburgh Edinburgh University Press, 2017.

McCullen, Jim. *Control Your Day: A New Approach to Email and Time Management Using Microsoft® Outlook and the concepts of Getting Things Done®.* Brookfield, WI: CreateSpace, 2013.

McDonald, Nora. "Distributed Leadership in OSS: Proceedings of the 18th International Conference on Supporting Group Work." New York: ACM Digital Library, 2014.

McGann, Jim. "Index Engines Finds More Dirt on Nuix's 'Cleansed' Enron Data Set." *Index Engines,* May 23, 2013. http://www.poweroverinformation.com/index-engines-finds-more-dirt-on-nuixs-cleansed-enron-data-set/.

McGill, Andrew. "Why White People Don't Use White Emoji." *The Atlantic*, May 9, 2016.

McLean, Bethany, and Peter Elkind. *The Smartest Guys in the Room: The Amazing Rise and Scandalous Fall of Enron*. New York: Portfolio Trade, 2003.

McLaughlin, Bryan, and Timothy Macafee. Becoming a PresidentialCandidate: Social Media Following and Politician Identification, *Mass Communication and Society* 22, no. 5 (2019): 584–603.

McMorris-Santoro, Evan. "Former Administration Aides: Expectation Was Disclosure for Personal Email." *BuzzFeed*, March 6, 2015. http://www.buzzfeed.com/evanmcsan/obama-administration-private-email#.vqG5nq2yn.

McQueen, Rob. "The US Postal Service in the Technological Climate." *The Tech*, March 20, 2012. http://tech.mit.edu/V132/N13/email.html.

McRobbie, Angela. *The Aftermath of Feminism*. London: SAGE, 2008.

Menke, Richard. *Telegraphic Realism: Victorian Fiction and Other Information Systems*. Stanford, CA: Stanford University Press, 2008.

Merica, Dan. "Clinton Used iPad for Personal Email at State." CNN Politics, March 31, 2015. http://edition.cnn.com/2015/03/31/politics/hillary-clinton-ipad-e-mail-devices/.

Merrier, Patricia A., and Ruthann Dirks. 1997. "Student Attitudes Toward Written Oral and E-mail Communication." *Business Communication Quarterly* 60, no. 2 (1997): 89–99.

Michelson, Leslie P. "Recollections of a Mentor and Colleague of a 14-Year-Old, Who Invented Email in Newark, NJ." inventorofemail.com, n.d. http://www.inventorofemail.com/va_shiva_recollection_inventing_email_leslie_michelson.asp.

Michelson, Leslie P. "The Invention of Email." inventorofemail.com, n.d. http://historyofemail.com/invention-of-email.asp.

Millard, William B. "I Flamed Freud: A Case Study in Teletextual Incendarism." In *Internet Culture*, edited by David Porter. New York: Routledge, 1997.

Miller, Daniel. *Stuff*. Cambridge: Polity, 2010.

Miller, Daniel, and Heather A. Horst. "The Digital and the Human: A Prospectus for Digital Anthropology." In *Digital Anthropology*, edited by Heather A. Horst and Daniel Miller. London: Berg, 2012.

Miller, Zeke J. "Transcript: Everything Hillary Clinton Said on the Email Controversy." *Time*, March 11, 2015.

Milne, Esther. *Letters Postcards Email: Technologies of Presence*. London: Routledge, 2010.

Milner, Ryan M. *The World Made Meme Public Conversations and Participatory Media* Cambridge, MA: MIT Press, 2016.

Miltner, Kate. "From #Feels to Structure of Feeling: The Challenges of Defining 'Meme Culture." Festival of Memeology. *Culture Digitally*, October 29, 2015. http://culturedigitally.org/2015/10/memeology-festival-02-from-feels-to-structure-of-feeling-the-challenges-of-defining-meme-culture/.

"Miranda July: From the Outboxes of the Noteworthy." *NPR*, June 6, 2017. http://www.npr.org/2013/07/06/199254433/from-the-outboxes-of-the-noteworthy.

MIT Comparative Media Studies/Writing. "The Future of the Post Office." April 20, 2016.

Moncarz, Elisa, Raul Moncarz, Alejandra Cabello, and Benjamin Moncarz.. "The Rise and Collapse of Enron: Financial Innovation, Errors and Lessons." *Accounting and Management* (2006): 17–37.

Moran, Joe. *Reading the Everyday*. London: Routledge, 2005.

Morris, Errol. "Did My Brother Invent Email with Tom Van Vleck?" *New York Times*, June 19, 2011. http://opinionator.blogs.nytimes.com/2011/06/23/did-my-brother-invent-e-mail-with-tom-van-vleck-part-five/.

Morrissey, Ed. "New Story From Hillary Camp: Sure, We Read all 32,000 Deleted E-mails First." *Hot Air*, March 16, 2015.

Mueller, Rob "Exciting News: Fastmail Staff Purchase the Business from Opera," *Fastmail* (blog), September 25 2013. https://blog.fastmail.com/2013/09/25/exciting-news-fastmail-staff-purchase-the-business-from-opera/.

Mullin, Benjamin. "Bustle Owner Plans to Double Down on Gawker." *Wall Street Journal*, September 17, 2018. https://www.wsj.com/articles/bustle-owner-plans-to-double-down-on-gawker-1537211032.

Mullin, Joe, and Cyrus Farivar. "History by Lawsuit: After Gawker's Demise, the "Inventor of E-mail" Targets Techdirt." *Ars Technica*, June 13 2017. https://arstechnica.com/tech-policy/2017/06/shivas-war-one-mans-quest-to-convince-the-world-that-he-invented-e-mail/.

Murphy, Patrick. "Locating Media Ethnography." In *The Handbook of Media Audiences*, edited by Virginia Nightingale. Oxford: Wiley-Blackwell, 2011.

Murray, Denise E. "CMC: A Report on the Nature and Evolution of On-Line E-Messages." *English Today* 6, no. 3 (1990): 42–46.

Nafus, Dawn and Ken Anderson. "Writing on Walls: The Materiality of Social Memory in Corporate Research." In *Ethnography and the Corporate Encounter: Reflections on Research in and of Corporations*, edited by Melissa Cefkin. New York: Berghahn Books, 2010.

Nanos, Janelle. "Return to Sender." *Boston Magazine*, June 2012.

Nantz, Karen, and Cynthia Drexel. 1995. "Incorporating Electronic Mail into the Business Communication Course." *Business Communication Quarterly* 58, no. 3 (1995): 45–51.

"National Archives Releases State Department Letter Re: Email Recordkeeping." 2015 press releases. National Archives, https://www.archives.gov/press/press-releases/2015/nr15-65.html.

National Research Council. *Review of Electronic Mail Service Systems Planning for the US Postal Service*. Washington DC: National Academies Press, 1981.

Neilson, Brett and Ned Rossiter. "From Precarity to Precariousness and Back Again: Labour, Life and Unstable Networks." *The Fibreculture Journal* 5 (2005).

Nelson, Ted. "Mail Chauvinism: The Magicians, the Snark and the Camel." *Creative Computing* 7, no. 11 (1981).

nettime Mailing Lists. Accessed May 28, 2017. http://www.nettime.org/.

"New EDRM Enron Email Data Set." EDRM Duke Law. Accessed June 8, 2017. http://www.edrm.net/resources/data-sets/edrm-enron-email-data-set.

Newman, Janet. "Bending Bureaucracy: Leadership and Multi-Level Governance." In *The Values of Bureaucracy*, edited by Paul du Gay. Oxford: Oxford University Press, 2005.

Nightingale, Virginia. "Media Ethnography and the Disappearance of Communication Theory." *Media International Australia* 145, no. 1 (2012).

Nippert-Eng, Christena E. *Islands of privacy* Chicago: University of Chicago Press, 2010.

Norberg, Arthur, and Judy O'Neill Freedman. *Transforming Computer Technology: Information Processing for the Pentagon, 1962–1986*. Baltimore: Johns Hopkins University Press, 1996.

Novak, Petra Kralj, Jasmina Smailović, Borut Sluban, and Igor Mozetič. "Sentiment of Emojis." *PLOS ONE* 10, no. 12 (2015): 1–22.

Offit, Avodah. *Virtual Love: A Novel*. New York: Simon & Schuster, 1994.

"Okta's Businesses @ Work Report." Accessed May 5, 2017. https://www.okta.com/Businesses-At-Work/2015-08/.

"Opera Acquires E-mail Service Fastmail.fm." *Hacker News*, May 1, 2010. https://news.ycombinator.com/item?id=1307649.

"Oral History of Raymond 'Ray' Tomlinson." Interviewed by Marc Weber and Gardner Hendrie. Cambridge, Massachusetts. Computer History Museum, 2009.

"Our Shadow Selves from Canada." *Dear You*. Accessed June 11, 2017. http://www.dearyouartproject.com/.

Ovans, Andrea. "Can E-mail Deliver the Message?" *Datamation* 38, no. 11 (1992).

Padlipsky, Mike. "And They Argued All Night ..." In *The ARPANET Sourcebook: The Unpublished Foundations of the Internet*, edited by Peter H. Salus (Charlottesville, VA: Peer-to-Peer Communications LLC, 2008).

Palumbo, Antonino, and Alan Scott. "Bureaucracy, Open Access and Social Pluralism: Returning the Common to the Goose." In *The Values of Bureaucracy*, edited by Paul du Gay. Oxford: Oxford University Press, 2005.

Parikka, Jussi. "Operative Media Archaeology: Wolfgang Ernst's Materialist Media Diagrammatics." *Theory, Culture & Society* 28, no. 5 (2011).

Parikka, Jussi. *What is Media Archaeology?* Cambridge: Polity, 2012.

Parikka, Jussi, and Tony Sampson, eds. *The Spam Book: On Viruses, Porn, and Other Anomalies from the Dark Side of Digital Culture*. Cresskill, NJ: Hampton Press, 2009.

Park, Hee Sun, Hye Eun Lee, Jeong An Song. "I Am Sorry to Send You SPAM': Cross-Cultural Differences in Use of Apologies in Email Advertising in Korea and US" *Human Communication Research* 31, no. 3 (2005): 365–398.

Partridge, Craig. "The Technical Development of Internet Email." *IEEE Annals of the History of Computing* 30, no. 2 (2008).

Pavalanathan, Umashanthi, and Jacob Eisenstein. "More Emojis, Less :) The Competition for Paralinguistic Function in Microblog Writing." *First Monday* 21, no. 11 (2016).

Perry, Ruth. *Women, Letters, and the Novel*. New York: AMS Press, 1980.

Pexton, Patrick B. "Origins of E-mail: My Mea Culpa." *Washington Post*, March 1, 2012, https://www.washingtonpost.com/blogs/omblog/post/origins-of-e-mail-my-mea-culpa/2012/03/01/gIQAiOD5kR_blog.html.

Pexton, Patrick B. "Reader Meter: Who Really Invented E-mail?" *Washington Post*, February 24, 2012, https://www.washingtonpost.com/blogs/omblog/post/reader-meter-who-really-invented-e-mail/2012/02/24/gIQAHZugYR_blog.html.

Pogran, Ken, John Vittal, Dave Crocker, and Austin Henderson. "Proposed Official Standard for the Format of ARPA Network Messages." RFC 724. May 12, 1977. https://doi.org/10.17487/RFC0724.

"Post Secret." Accessed June 11, 2017. http://postsecret.com/.

Postel, Jon. "Comments on RFC 724." Post to the MSGGRP discussion list, May 23, 1977.

Postel, Jon. "File Transfer Protocol Specification." 1980. https://doi.org/10.17487/RFC0765.

Postel, Jon. "Simple Mail Transfer Protocol." STD 10, RFC 821, August 1982. https://doi.org/10. 17487/RFC0821.

Postigo, Hector. "America Online Volunteers: Lessons from an Early Co-production Community." *International Journal of Cultural Studies* 12, no. 5 (2009): 451–469.

Priebus, Reince. "Reince Priebus: 53 Questions I Have for Hillary Clinton." *Independent Journal Review*. Accessed June 11, 2017. http://journal.ijreview.com/2015/03/242744-reince-priebus-53-questions-hillary-clinton/.

Pynchon, Thomas. *The Crying of Lot 49*. Philadelphia: Lippincott, 1966.

Raghavan, Anita, Kathryn Kranhold, and Alexei Barrionuevo. "How Enron Bosses Created A Culture of Pushing Limits." *Wall Street Journal*, August 26, 2002. https://www.wsj.com/articles/SB1030320285540885115.

Rasmussen, Terje. *Personal Media and Everyday Life: A Networked Lifeworld*. London: Palgrave Macmillan, 2014.

Ravenscraft, Eric. "Disable Automatic Image Loading In Gmail To Improve Security." *Lifehacker*, December 15, 2013.

Reutera, Christian, Thomas Ludwiga, Marc-André Kaufhold, and Thomas Spielhofer. "Emergency Services' Attitudes towards Social Media: A Quantitative and Qualitative Survey across Europe." *International Journal of Human-Computer Studies* 95 (November 2016): 96–111.

"Response to Under Secretary of State for Management, Patrick Kennedy from Hillary Clinton's Representative, Cheryl Mills, December 5, 2014." National Archives 2015 press releases.

RFC Editor. "About Us." https://www.rfc-editor.org/about/.

Riddell, Alan, ed. *Typewriter Art*. London: London Magazine Editions, 1975.

Riedy, Marian K., and Joseph H. Wen. "Electronic Surveillance of Internet Access in the American Workplace: Implications for Management." *Information & Communications Technology Law* 19, no. 1 (2010): 87–99.

Rivière, Carole Anne, and Christian Licoppe. "From Voice to Text: Continuity and Change in the Use of Mobile Phones in France and Japan." In *The Inside Text*, edited by Richard Taylor, Leysia Ann Palen, and A. Harper. Netherlands: Springer, 2006.

Roberts, Dan. "Hillary Clinton Breaks Media Silence and Insists: 'I Want Those Emails Out.'" *Guardian*, May 19, 2015. http://www.theguardian.com/us-news/2015/may/19/hillary-clinton-emails-media-iowa.

Roberts, Dan. "Hillary Clinton Foists Private Email Release Delays on Obama Administration." *Guardian*, July 30, 2015. http://www.theguardian.com/us-news/2015/jul/30/hillary-clinton-emails-release-delay-state-department.

Robinson, Victoria. "Reconceptualising the Mundane and the Extraordinary: A Lens through Which to Explore Transformation within Women's Everyday Footwear Practices." *Sociology* 49, no. 5 (2015): 903–918.

Rodriguez, Noelie Maria. "Transcending Bureaucracy: Feminist Politics at a Shelter for Battered Women." *Gender & Society* 2, no. 2 (1988): 214–227.

Rogers, Juan D., and Gordon Kingsley. "Denying Public Value: The Role of the Public Sector in Accounts of the Development of the Internet." *Journal of Public Administration Research and Theory* 14, no. 3 (2004).

Rooksby, Emma. *E-Mail and Ethics: Style and Ethical Relations in Computer-Mediated Communication.* London: Routledge, 2002.

Rosenberg, Scott. "Shut Down Your Office. You Now Work in Slack." *Backchannel,* May 8, 2015. https://backchannel.com/shut-down-your-office-you-now-work-in-slack-fa83cb7cce6c#.uwm9wd887

Rossiter, Ned. "Logistical Worlds." *Cultural Studies Review* 20, no. 1 (2014): 53–76.

Rotunno, Laura. *Postal Plots in British Fiction, 1840–1898: Readdressing Correspondence in Victorian Culture.* New York: Palgrave Macmillan, 2013.

Russell, Andrew L. "OSI: The Internet That Wasn't." *IEEE Spectrum,* July 30, 2013.

Russell, Andrew L. *Open Standards and the Digital Age: History, Ideology, and Networks.* New York: Cambridge University Press, 2014.

Sackner, Marvin, and Ruth Sackner. *The Art of Typewriting.* London: Thames & Hudson, 2015.

Salahub, Christopher, and Wayne Oldford, "About 'Her Emails.'" *Significance* 15, no. 3 (June 2018): 34–37. https://doi.org/10.1111/j.1740-9713.2018.01148.x.

Salus, Peter, H. *Casting the Net: From ARPANET to Internet and Beyond* Reading, MA: Addison-Wesley, 1995.

Samanage. "Is Work Email Disrupting the Personal Lives of US Employees?" *Samanage.* March 30, 2016. https://www.samanage.com/company/news/samanage-survey-shows-employees-squander-more-than-a-month-each-year/.

Samoriski, Jan, and John Huffman. "Electronic Mail, Privacy, and the Electronic Communications Privacy Act of 1986: Technology in Search of Law." *Journal of Broadcasting & Electronic Media* 40, no 1 (1996).

Savage, Mike, and Anne Witz. *Gender and Bureaucracy.* London: Blackwell, 1992.

Schenker, Brett. "The 2017 Nonprofit Email Deliverability Study: How Much Does Spam Hurt Online Fundraising?" EveryAction, 2017.

Schmidt, Michael S. "Hillary Clinton Used Personal Email Account at State Dept., Possibly Breaking Rules." *New York Times,* March 2, 2015. http://www.nytimes.com/2015/03/03/us/politics/hillary-clintons-use-of-private-email-at-state-department-raises-flags.html.

Scholz, Trebor, ed. *Digital Labour: The Internet as Playground and Factory*. London: Routledge, 2013.

Scott, Steven. "Paper Can Be Clean and Green, Industry Claims." *The Australian Financial Review*, September 20, 2010.

Shaller, Douglas. "E-mail, the Internet, and Other Legal and Ethical NIGHTMARES." *Strategic Finance* 82, no. 2 (2000): 48–52.

Shapiro, Norman Z., and Robert H. Anderson. *Toward an Ethics and Etiquette for Electronic Mail*. Santa Monica, CA: Rand, 1985.

Sharma, Mahesh. "Kill off Email to Boost Productivity." *The Age*, April 11, 2014. http://www.theage.com.au/digital-life/digital-life-news/kill-off-email-to-boost-productivity-20140411-zqtik.html.

Sherblom, John. "Direction, Function, and Signature in Electronic Mail." *International Journal of Business Communication* 25, no. 4 (1988): 39–54.

Shetty, Jitesh, and Jafar Adibi. "The Enron Email Dataset Database Schema and Brief Statistical Report." *Information Sciences Institute Technical Report*. Los Angeles: University of Southern California, 2004.

Shihab, Emad, Nicholas Bettenburg, Bram Adams, and Ahmed E. Hassan. "On the Central Role of Mailing Lists in Open Source Projects: An Exploratory Study." In *New Frontiers in Artificial Intelligence*, edited by Yohei Murakami and Eric McCready Kumiyo Nakakoji. Berlin: Springer, 2010.

Shimmin, Bradley, *Effective E-mail Clearly Explained: File Transfer, Security, and Interoperability*. Open Library: AP Professional, 1997.

Shipley, David. *SEND: Why People Email So Badly and How to Do It Better*. Visalia, CA: Vintage, 2010.

Simon, Sunka. *Mail-Orders: The Fiction of Letters in Postmodern Culture*. Albany: State University of New York, 2002.

Smith, Lauren. "Email Client Market Share: Where People Opened in 2013." *Litmus* (blog), January 16, 2014. https://litmus.com/blog/email-client-market-share-where-people-opened-in-2013.

Smith, William, and Filiz Tabak. "Monitoring Employee E-mails: Is There Any Room for Privacy?" *Academy of Management Perspectives* 23, no. 4 (2009): 33–48.

Smithsonian Institute. "Statement from the National Museum of American History: Collection of Materials from V.A. Shiva Ayyadurai," 2012.

Snyder, Jason L. 2010. "E-mail Privacy in the Workplace: A Boundary Regulation Perspective." *Journal of Business Communication* 47, no. 3 (2010): 266–294.

Sondheim, Alan. "Note to Jon Marshall about Living on Cybermind." *Cybermind,* January 2, 2017. https://listserv.wvu.edu/cgi-bin/wa?A0=CYBERMIND.

Sorkin, Andrew Ross. "Peter Thiel, Tech Billionaire, Reveals Secret War with Gawker." *New York Times,* May 25, 2016. http://www.nytimes.com/2016/05/26/business/dealbook/peter-thiel-tech-billionaire-reveals-secret-war-with-gawker.html.

Sproull, Lee, and Sara Kiesler. "Reducing Social Context Cues: Electronic Mail in Organizational Communication." *Management Science* 32, no. 11 (1986): 1492–1512.

Star, Susan Leigh. "This is Not a Boundary Object: Reflections on the Origin of a Concept." *Science, Technology, & Human Values* 35, no 5 (2010): 601–617.

Stark, Luke, and Kate Crawford. "The Conservatism of Emoji: Work, Affect, and Communication." *Social Media + Society* 1, no. 2 (2015).

"Small Scale Performance Propositions." *Realtime,* accessed June 11, 2017. http://www.realtimearts.net/studio-artist/small-scale-performance-propositions.

"State Department Response to National Archives' Letter Re: State Department Emails." National Archives press releases 2015, March 3, 2015.

"State Department's Response to National Archives Chief Records Officer, Paul M. Wester's March 3, 2015, Letter about Email Practices of Former Secretaries of State." National Archives.

Statista. "Spam: Share of Global Email Traffic 2014–2016." https://www.statista.com/statistics/420391/spam-email-traffic-share/.

Statistic Brain. "Paper Use Statistics." October 29, 2016. http://www.statisticbrain.com/paper-use-statistics/.

Steadman, Carl. "Two Solitudes." *InterText* 5, no. 1 (1995).

Steen, Margaret. "Legal Pitfalls of E-mail." *InfoWorld* 21, no. 27 (1999): 65–66.

Stephens, Keri K. *Negotiating Control: Organizations and Mobile Communication.* New York: Oxford University Press, 2018.

Stern, Remy. "I Had a One-Night Stand with Christine O'Donnell." *Gawker,* October 28, 2010. http://gawker.com/5674353/i-had-a-one-night-stand-with-christine-odonnell.

Stivers, Camilla. *Gender Images in Public Administration: Legitimacy and the Administrative State.* 2nd ed. Thousand Oaks, CA: Sage, 2002.

Stromberg, Joseph. "A Piece of Email History Comes to the American History Museum." *Smithsonian.com,* February 22, 2012.

Styler, Will. "The EnronSent Corpus. Technical Report '01–2011." University of Colorado at Boulder Institute of Cognitive Science, 2011.

Sutton, Selina Jeanne. "Emoji Are Becoming More Inclusive, but Not Necessarily More Representative." *The Conversation*, February 8, 2019. https://theconversation.com/emoji-are-becoming-more-inclusive-but-not-necessarily-more-representative-111388.

Swalwell, Melanie. "Questions about the Usefulness of Microcomputers in 1980s Australia." *Media International Australia* 143 (2012): 63–77.

Taylor-Smith, Ella, and Colin Smith. "Non-public eParticipation in Social Media Spaces." *SMSociety '16 Proceedings of the 7th 2016 International Conference on Social Media & Society*. No. 3 (July 8, 2016): 1–8.

"Task Force on Technical Approaches to Email Archives." Accessed June 11, 2017. http://www.emailarchivestaskforce.org/about/task-force-charge/.

Tekla, Perry. "E-mail Pervasive and Persuasive: The Net of Networks Now Embracing the Globe Is Bypassing Corporate and Other Hierarchies." (Special Report/Electronic Mail) *IEEE Spectrum* 29, no. 10 (October 1992: 22–23.

Terranova, Tiziana. "Free Labor: Producing Culture for the Digital Economy." *Social Text* 18, no. 2 (2000): 33–34.

The Association of Internet Researchers. Air-L Archives. http://listserv.aoir.org/pipermail/air-l-aoir.org/.

"The Browning Letters." Baylor University. Accessed June 11, 2017. http://digitalcollections.baylor.edu/cdm/landingpage/collection/ab-letters.

"The Collaborative Electronic Records Project." The Rockefeller Archive Center. Accessed June 12, 2017. http://siarchives.si.edu/cerp/index.htm.

The Environmental Paper Assessment Tool. https://www.epat.org/.

"The Fall of Enron." *NPR*. Accessed June 4, 2017. http://www.npr.org/news/specials/enron/.

The International Consortium of Investigative Journalists. "Giant Leak of Offshore Financial Records Exposes Global Array of Crime and Corruption." April 3, 2016.

The Internet Engineering Task Force. https://www.ietf.org/list/.

The Mail Archive. https://www.mail-archive.com/.

"The Sackner Archive of Concrete and Visual Poetry." Accessed June 11, 2017. http://ww3.rediscov.com/sacknerarchives/Welcome.aspx?518201785314.

The Wikimedia-l Archives. Wikimedia Foundation Mailing List. https://lists.wikimedia.org/pipermail/wikimedia-l/.

"Thierry Breton, Chairman and CEO of Atos." Atos website. http://atos.net/en-us/home/we-are/zero-email.html.

Thomas, Katie-Louise. *Postal Pleasures: Sex, Scandal, and Victorian Letters*. New York: Oxford University Press, 2012.

Thompsen, Philip A. "What's Fueling the Flames in Cyberspace? A Social Influence Model." In *Communication and Cyberspace: Social Interaction in an Electronic Environment*, edited by Lance Strate, Ron L. Jacobson, and Stephanie Gibson. Cresskill, NJ: Hampton, 1996.

Thompsen, Philip A., and Davis A. Foulger. "Effects of Pictographs and Quoting on Flaming in Electronic Mail." *Computers in Human Behavior* 12, no. 2 (1996): 225–243.

Tronto, Joan C. *Caring Democracy Markets, Equality, and Justice*. New York: NYU Press, 2013.

Trotter, J. K. "Univision Executives Vote to Delete Six Gawker Media Posts." *Gizmodo*, September 10, 2016.

Tullett, Barrie. *Typewriter Art: A Modern Anthology*. London: Laurence King Publishing, 2014.

Turco, Catherine J. *The Conversational Firm: Rethinking Bureaucracy in the Age of Social Media*. New York: Columbia University Press, 2016.

Turnage, Anna K. "Email Flaming Behaviors and Organizational Conflict." *Journal of Computer-Mediated Communication* 13 (2008): 43–59.

Tutt, Paige. "Apple's New Diverse Emoji Are Even More Problematic Than Before." *Washington Post*, April 10, 2015. https://www.washingtonpost.com/posteverything/wp/2015/04/10/how-apples-new-multicultural-emojis-are-more-racist-than-before/.

"Unicode Emoji, Technical Report #51." In *Unicode*, November 22, 2016.

Unify. "The Way We Work: 9,000 Knowledge Workers Share Their Insight on the Jobs They Do and the Workplace of the Future." Accessed May 5, 2017. http://www.economyup.it/upload/images/11_2016/161122122232.pdf.

United States Copyright Office. "Computer Program for Electronic Mail System." 1981.

United States District Court District of Massachusetts. SHIVA AYYADURAI Plaintiff v. FLOOR64, INC. d/b/a TECHDIRT, MICHAEL DAVID MASNICK, LEIGH BEADON, and DOES 1–20 [reference to dates is on page 4], Civil Action No. 17–10011-FDS.

"Updated: The Facts About Hillary Clinton's Emails." *The Briefing*, December 11, 2015. https://www.hillaryclinton.com/briefing/factsheets/2015/07/13/email-facts/.

US Department of Justice (Antitrust Division), "Proposed Revisions in the Comprehensive Standards for Permissible Private Carriage of Letters," March 13, 1979, 16.

Uricchio, William. "History and its Shadow: Thinking About the Contours of Absence." *Screen* 55, no. 1 (2014): 119–127.

van der Nagel, Emily. "'Networks That Work Too Well': Intervening in Algorithmic Connections." *Media International Australia* 168, no. 1 (2018): 81–92.

van Dijck, José. *The Culture of Connectivity: A Critical History of Social Media*. Oxford: Oxford University Press, 2013.

van Dijck, José. "Datafication, Dataism and Dataveillance: Big Data Between Scientific Paradigm and Ideology." *Surveillance & Society* 12, no. 2 (2014): 197–208.

Van Vleck, Tom. "Dead Media Beat: The History of Electronic Mail." *Wired*, April 28, 2013.

Van Vleck, Tom. "Electronic Mail and Text Messaging in CTSS, 1965–1973." *IEEE Annals of the History of Computing* 34, no. 1 (2012): 46.

Van Vleck, Tom. "The History of Electronic Mail." n.d. http://multicians.org/thvv/mail-history.html.

Victoria 19th-Century British Culture & Society. "Indiana University Mailing List." https://list.indiana.edu/sympa/arc/victoria.

Victoria Research Web. "The Electronic Conference for Victorian Studies." Accessed May 28, 2017. http://victorianresearch.org/discussion.html.

Vinh-Doyle, William P. "Appraising Email (Using Digital forensics): Techniques and Challenges." *Archives and Manuscripts* 45, no. 1 (2017).

Volkova, Svitlana, Theresa Wilson, and David Yarowsky. "Exploring Demographic Language Variations to Improve Multilingual Sentiment Analysis in Social Media." Proceedings of the 2013 Conference on Empirical Methods in Natural Language Processing (2013): 1815–1827.

Voorhees, Josh. "A Crystal Clear Explanation of Hillary's Confusing Email Scandal." *The Slatest* (blog). August 20, 2015. http://www.slate.com/blogs/the_slatest/2015/08/20hillary_clinton_email_scandal_explained.html.

Vrooman, Steven S. "The Art of Invective: Performing Identity in Cyberspace." *New Media & Society* 4, no. 1 (2002): 51–70.

Wajcman, Judy. *Pressed for Time: The Acceleration of Life in Digital Capitalism*. Chicago: University of Chicago Press: 2015.

Walden, Dave. "A Fresh Tissue of Lies." *SIGCIS*, March 24, 2016. http://lists.sigcis.org/pipermail/members-sigcis.org/2016-March/000134.html.

Waldow, D. J., and Jason Fells. *The Rebel's Guide to Email Marketing: Grow Your List, Break the Rules, and Win*. Indianapolis: Que Publishing, 2012.

Walker, Tom. "The Evolution of Printer Technology: Then and Now." *Cartridgesave.co.uk* April 15, 2008. http://www.cartridgesave.co.uk/news/the-evolution-of-printer-technology-then-and-now/.

Waller, Vivienne. "This Big Hi-tech Thing': Gender and the Internet at Home in the 1990s." *Media International Australia* 143 (2012): 78–88.

Walshe, Shushannah, and Liz Kreutz. "Hillary Clinton's Deleted Emails Were Individually Reviewed After All, Spokesman Says." *ABC News*, March 15, 2015. http://abcnews.go.com/Politics/hillary-clintons-deleted-emails-individually-reviewed-spokesman/story?id=29654638.

Walt, Vivienne. "France's 'Right to Disconnect' Law Isn't All It's Cracked Up to Be." *Time*, January 5, 2017.

Watkins, Jonathan, and René Denizot. *On Kawara*. London: Phaidon, 2002.

Watson, Nicola J. *Revolution and the Form of the British novel, 1790–1825: Intercepted Letters, Interrupted Seductions*. Oxford: Clarendon Press, 2001.

Waugh, Andrew. "Email—A Bellwether Records System." *Archives and Manuscripts* 42, no. 2 (2014).

"We Think Alone: A Project by Miranda July." https://www.magasin3.com/en/pressrelease/we-think-alone-a-project-by-miranda-july-2/.

"What is Immersion?" *Immersion*, accessed June 11, 2017 https://immersion.media.mit.edu/.

"What is Mailart?" Mailart 365, accessed June 11, 2017. http://www.mailart365.com/what-is-mailart-365/.

Weeks, Kathi. "Life Within and Against Work: Affective Labor, Feminist Critique, and Post-Fordist Politics." *ephemera* 7, no. 1 (2007): 233–249.

Welch, Chuck, ed. *Eternal Network: A Mail Art Anthology*. Calgary: University of Calgary Press, 1994.

Wemple, Erik. "Conde Nast Exec Story: Gawker is Keeping Its Sleaze Game in Shape." *Washington Post*, July 17, 2015. https://www.washingtonpost.com/blogs/erik-wemple/wp/2015/07/17/conde-nast-exec-story-gawker-is-keeping-its-sleaze-game-in-shape/.

Werbach, Kevin. "Death by Spam: The E-mail You Know and Love Is About to Vanish." *Slate*, November 18, 2002.

Whittaker, Steve, Victoria Bellotti, and Paul Moody. "Introduction to This Special Issue on Revisiting and Reinventing E-Mail." *Human Computer Interaction* 20 (2005): 1–9.

"WikiLeaks: From Popular Culture to Political Economy." *International Journal of Communication* 8 (2014).

"WikiLeaks List: Most Damaging Emails About DNC, Clinton, & Bernie." *Heavy*. July 24, 2016. http://heavy.com/news/2016/07/wikileaks-emails-clinton-bernie-list-directory-photos-most-damaging-worst-rhode-island-delegate-fec-jvf/.

WikiLeaks. "Sony." Accessed June 4, 2017. https://wikileaks.org/sony/press/.

Wilkes, Maurice V. "Networks, Email and Fax." *Communications of the ACM* 33, no. 6 (1990): 631–633.

Wilson, Garnett, and Wolfgang Banzhaf. "Discovery of Email Communication Networks from the Enron Corpus with a Genetic Algorithm using Social Network Analysis." IEEE Congress on Evolutionary Computation Conference, Trondheim, Norway, May 2009.

Wilson, Jason, Glen Fuller, and Christian McCrea, eds. "Trolls and The Negative Space of the Internet." *The Fibreculture Journal* 22 (2013).

Wohlsen, Marcus. "The Next Big Thing You Missed: Email's About to Die, Argues Facebook Co-Founder." *Wired*, January 1, 2014

Wong, Kristina. "Happy New Year! State Drops 5,500 Pages of Clinton Emails." *The Hill*, December 31, 2015. http://thehill.com/policy/national-security/264527-feds-release-new-batch-of-clinton-emails-on-new-years-eve.

Wood, Pat, William Massey, Linda Breathitt, and Nora Mead Brownell (Commissioners). Fact-Finding Investigation of Potential Manipulation of Electric and Natural Gas Prices. Federal Energy Regulatory Commission, February 13, 2002.

Vorberg, Laura and Anna Zeitler. "'This Is (Not) Entertainment!': Media Constructions of Political Scandal Discourses in the 2016 US Presidential Election. *Media, Culture & Society* 41, no. 4 (2019): 417–432.

Yahoo! Groups. "DNA-NEWBIE." The International Society of Genetic Genealogy. Accessed May 28, 2017. https://groups.yahoo.com/neo/groups/DNA-NEWBIE/info.

Yahoo! Groups. "Crossfire—Tabletop Miniature Wargaming." Accessed May 28, 2017. https://groups.yahoo.com/neo/groups/Crossfire-WWII/info.

Yates, JoAnne. *Control through Communication: The Rise of System in American Management*. Baltimore: Johns Hopkins University Press, 1989.

Yates, JoAnne. "The Emergence of the Memo as a Managerial Genre: Methods, Approach" *Management Communication Quarterly* 2, no. 4 (1989).

Yates, Simeon J. "Oral and Written Linguistic Aspects of Computer Conferencing: A Corpus Based Study." In *Computer-Mediated Communication: Linguistic, Social and Cross-Cultural Perspectives*, edited by Susan C. Herring. Amsterdam: Benjamins, 1996.

Young, Tim. "Social Networks Spur the Demise of Email in the Workplace." *Socialcast Blog*, June 7, 2010. http://blog.socialcast.com/social-networks-spur-the-demise-of-email-in-the-workplace/.

Zhou, Yingjie. "Mining Organizational Emails for Social Networks with Application to Enron Corpus (PhD dissertation)." Rensselaer Polytechnic Institute, 2008.

Zhu, Wuhan. "Polite Requestive Strategies in Emails: An Investigation of Pragmatic Competence of Chinese EFL Learners." *RELC Journal* 43 (2012): 217–238.

Zielinski, Siegfried. *Deep Time of the Media*. Cambridge, MA: MIT Press, 2006.

Zilberman, Polina, Gilad Katz, Asaf Shabtai, and Yuval Elovici. "Analyzing Group E-mail Exchange to Detect Data Leakage." *Journal of the American Society for Information Science and Technology* 64, no. 9 (2013): 1780–1790.

Index

Printed in the United States
by Baker & Taylor Publisher Services